The Bioethics of Space Exploration

The Bioethics of Space Exploration

Human Enhancement and Gene Editing in Future Space Missions

Konrad Szocik

OXFORD
UNIVERSITY PRESS

OXFORD
UNIVERSITY PRESS

Oxford University Press is a department of the University of Oxford. It furthers
the University's objective of excellence in research, scholarship, and education
by publishing worldwide. Oxford is a registered trade mark of Oxford University
Press in the UK and certain other countries.

Published in the United States of America by Oxford University Press
198 Madison Avenue, New York, NY 10016, United States of America.

Library of Congress Cataloging-in-Publication Data
Names: Szocik, Konrad, author.
Title: The bioethics of space exploration : human enhancement and
gene editing in future space missions / Konrad Szocik.
Description: New York, NY, United States of America : Oxford University Press, 2023. |
Includes bibliographical references and index.
Identifiers: LCCN 2022030352 (print) |
LCCN 2022030353 (ebook) | ISBN 9780197628478 (hardback) | ISBN 9780197628492 (epub)
Subjects: LCSH: Astronauts—Physiology. | Manned space flight—Moral and ethical aspects. |
Manned space flight—Health aspects. | Gene editing—Moral and ethical aspects. |
Performance technology—Moral and ethical aspects. |
Human body and technology—Moral and ethical aspects.
Classification: LCC TL856 .S96 2022 (print) |
LCC TL856 (ebook) | DDC 629.4—dc23/eng/20220822
LC record available at https://lccn.loc.gov/2022030352
LC ebook record available at https://lccn.loc.gov/2022030353

DOI: 10.1093/oso/9780197628478.001.0001

Printed by Sheridan Books, Inc., United States of America

For my wife, Rakhat Abylkasymova, my greatest love

Contents

Acknowledgments

I carried out the work on this book during my fellowship at the Interdisciplinary Center for Bioethics at Yale University during the 2021/2022 academic year. My gratitude goes to Stephen Latham, director of the Center, for the invitation and opportunity to conduct the work on the book, as well as to Lori Bruce.

I thank all those who shared their comments and feedback on my book: Rakhat Abylkasymova, Mark Shelhamer, Jacob Haqq-Misra, Ted Peters, Francis Cucinotta, Stephen Latham, Jonathan Anomaly, Koji Tachibana, Scott Solomon, Chris Impey, Henry Hertzfeld, Michael J. Reiss, Erik Persson, Tony Milligan, Martin Braddock, and Gonzalo Munévar. I should emphasize that not every person mentioned shared my views on the bioethical issues discussed or my vision of a future human presence in space. These are by nature often controversial and debatable issues. I thank the Yale University Library for guaranteeing prompt and always reliable access to all sources. Many thanks to Jeremy Lewis, Zoe Barham, and the entire Oxford University Press team involved in the preparation of the book. Finally, I am very grateful to the three anonymous reviewers for their very valuable comments. I also thank the Polish National Agency for Academic Exchange, which provided financial support for my one-year research stay at Yale University under the Bekker Fellowship (3rd edition) (Decision No. PPN/BEK/2020/1/00012/DEC/1).

Finally, I thank my wife, Rakhat Abylkasymova, for her patience and support as well as her very valuable guidance and advice. Without her, this book would probably never have been written.

1

The Place of Space Bioethics in the Philosophy and Ethics of Space Missions

Introduction

This is a book about the future of humanity in space, both the near future (missions to Mars and space mining) and the farther future that may never come (space colonization and space refuge), and how we should morally evaluate the possibilities, plans, scenarios, but also the consequences of various scenarios we pursue in the future with respect to space exploration.[1] Central to my thinking about the future of humans in space is the concept of human enhancement, which should be seen as an integral part of the larger whole of caring for our good future (Peters 2019a).

Thinking about the future in moral terms is particularly challenging for at least three reasons. First, we must anticipate all scenarios that might occur, even those that are extremely unrealistic. Second, even if we can predict all scenarios, we must have some way to determine which of these scenarios is most likely to occur, and at least under what conditions. Third, and finally, even if we somehow gain certainty about which scenario will come to pass, and it is, in our view, the worst-case scenario, or at least not the best-case scenario, we must prove why we think it is morally wrong. If we prove it, do we necessarily face the necessity of preventing this scenario? The answer is not that simple at all. Moral calculus changes over time and can be situational. In the context of future space missions, we can apply different moral categories to professional astronauts and others to commercial astronauts. Perhaps the risk of imminent planetary extinction will force us to flee Earth at all costs. Finally, advances in technology may open up entirely new possibilities. All of these can change the cost/benefit, which, in turn, can influence preferred moral principles.

How difficult, if at all possible, this is may be demonstrated by the following thought experiment. We agree that slavery is morally wrong. We also agree that genocide and depriving indigenous people of their land and freedom is morally wrong. We also agree that war is morally wrong. These are the things we want to avoid. Christopher Columbus probably did not think that the discovery of America would lead to these phenomena in the lands he discovered. At what point in our time machine should we stop and prevent the beginning of the sequence of events that led to slavery, the murder and deprivation of indigenous people of their land in the Americas, and finally the use of the atomic bomb against civilians for the only time in history? Is it enough to go back to the time just before Columbus was born, or perhaps to the time

The Bioethics of Space Exploration. Konrad Szocik, Oxford University Press. © Oxford University Press 2023. DOI: 10.1093/oso/9780197628478.003.0001

before ships were invented? But maybe we should go back to the source, that is, science, and prevent the development of mathematics and physics in ancient Greece and even earlier.

Someone might say that such thinking leads to paradoxes, but this is only an appearance. Let us rephrase our thought experiment and ask as follows: if you knew that achieving America's present status as a world power required the aforementioned phenomena as necessary conditions, would you, having a time machine at your disposal, have stopped this sequence of events? Or would you recognize that slavery and wars were the so-called spirit of the times, that, although morally wrong in themselves, they were accepted at a certain stage of human development, that, in other words, people once did not know that things could be different, did not quite know what they were allowed and what they were not? Perhaps, in a sense, this was the case if we look at the excellent distinctions made by Allen Buchanan and Rachel Powell in their book *The Evolution of Moral Progress*, where the authors point out that in the history of morality and culture, we have had periods of proper moralization (when we used to consider morally good and acceptable those acts that we today reject) and proper demoralization (behaviors in the past that were rejected, today morally acceptable and even desirable) (Buchanan and Powell 2018). Is slavery the "only" proper moralization that we have made in recent decades, correcting the mistakes of our ancestors? We can do similar thought experiments on all phenomena of the modern world that have negative consequences, like the Internet, climate change, computers. There are many negative social and criminal consequences associated with Internet and computer use, but being able to foresee them decades ago, should we conclude that humanity should not embark on a path of development leading to their invention and popularization?

This question need not sound at all as altruistic as it may seem, and it need not imply our concern for all the humanity that has come into existence and that has suffered in the centuries that have followed. This question, posed by each of us today, is very selfish in its nature. In its egoistic variant, it is as follows: Can I sacrifice myself, my coming into existence in the twentieth or twenty-first century, just to prevent the sequence of events of the past which led not only to wars and genocides but also made possible my coming into the world as this person with this and not that identity? This question implies many of the problems discussed by philosophers, such as the nonidentity problem and the idea of antinatalists that it is better for us never to have been born because, after all, since we never existed, we cannot know what we are missing by never coming into the world.

The search for the consequences of our actions and decisions that we make now is what I want to turn my attention to in relation to our future in space. I want to determine both what the possible scenarios for our future presence in space are, whether they are worth pursuing, and if so, how we should morally evaluate each of these options. I want to answer the question of whether every vision of humanity's future in space can be accepted, and whether there are scenarios that we should strongly reject on moral grounds. If there are such, then maybe we should stop the sequence of events right now that might lead to their realization. Should we pursue a space program that requires the application of radical forms of human enhancement, and is space exploration worth such severe modifications to human biology (Abylkasymova 2021)? Even given the moral uncertainty of the use of the atomic bomb, knowing how it led to a nuclear arms race, would it have been better to not use it at all, and to allow the continuation of a world war?

Even today, we do not know for sure whether these phenomena that are already happening (and that have set in motion a sequence of events leading, for example, to the establishment of a space colony at some point in the future) will definitely lead to the establishment of a space colony, and if so, whether the morally unacceptable option that we would like to avoid must unavoidably be realized in that colony when we think about Columbus's voyage to America and what happened to both the indigenous people and the future slaves. But the same may be true if we never leave Earth and, by forgoing space exploration, lead to an undesirable scenario on Earth in the future.

The axis for my consideration of what might happen in the future and what we should pursue or avoid contains two elements. The first is the aforementioned context of space missions. However, my considerations are perhaps even more about the Earth, how we think, how we make decisions, what our intuitions and moral rationales are, whether we are consistent in our thinking and moral judgment. The latter is a great challenge for us, because we often critically appraise one element, not noticing a given characteristic judged negatively in other phenomena that we accept. It is, after all, on Earth that a sequence of events is underway that will lead us to some scenarios in the future, including in space. What we are already doing on Earth today, and what decisions we will soon make about accepting what technologies and for what needs, will modify the sequence of events in morally acceptable, controversial, or unacceptable directions.

The second element central to my book is the concept of human enhancement through biomedicine, particularly gene editing. This narrows my discussion to the bioethical problem of moral acceptance of human enhancement for space missions. In this way I avoid considering other very important aspects of space mission ethics and philosophy that are currently being discussed by space philosophers and space ethicists (Green 2021).

Suffice it to say that we could bring about another total destruction of every newly exploited object in space, so it is not clear whether we have the right to think about space exploitation at all. Moreover, perhaps focusing our efforts on space exploration will reflect negatively on our concern for the Earth. Finally, a serious issue is that of debris in space, which at some point may be so abundant as to make it physically impossible to carry out any space operations (Kessler and Cour-Palais 1978), (Krag 2021).

Perhaps the colonization of space will involve scenarios that we morally disapprove of, but that will emerge with irresistible force. Perhaps we should avoid them by modifying the sequence of events today. Or maybe they just do not hide anything terrible in them. These are the guiding questions and the guiding objectives of this book. What may pose challenges and cause problems is neither gene editing nor space exploration itself, but a combination of the two, a constellation of events in which at some point they will have to occur together and lead to morally undesirable consequences. I am not saying that this must happen, however I am considering possible future scenarios in space in which this will be possible. I also point out that the same human enhancement that can be the source of moral problems can also be a tool for solving and preventing them.

These considerations are important for the Earth in the sense that if we recognize that certain biomedical procedures are controversial and we want to avoid them, and on the other hand we strive for space exploration, which may lead us to the colonization of space, then we may get new motivation to work on this "controversial" biotechnology because of the expected benefits. However, this does not change the fact that then the same technology can sometimes be used for the wrong purposes. Perhaps space missions will be the area where today's controversial human modification procedures will be widely applied. This is what I assume when discussing germline gene editing (GGE) for space missions.

In a sense, this is a book about humanity's future in the context of how its plans and ambitions are intertwined with the development of technologies it is entitled to, and which it is entitled to use for its own use and to improve its well-being, moral intuitions, and the long-term consequences of its choices. Perhaps that last element is the most important one—paying attention to the long-term consequences of our decisions and actions that we make today, which may direct the sequence of events toward something we would not want. Our presence in the cosmos can thus be seen as part of this larger whole, as a kind of case study.

It is at the same time a book about whether we should accept the unfettered development of science and technology, which, after all, is always dictated by the good and the well-being of humanity. It is a book about whether the end justifies the means, even if that end is the survival of our species.

Inevitably, I discuss the various moral dilemmas that arise today, and of which there will be even more as our capabilities for both space exploration and gene editing become more real and effective. In several of my articles (one with space philosopher Koji Tachibana) I posed the following dilemma, which I believe is inevitable. If we just accept that our future is in space and that we want to be present in space in a significant and meaningful way, we must choose one of two options: Either we accept gene editing on a fairly large scale for our adaptation to space conditions, or we give up our presence in favor of pursuing automated missions that require the development of highly advanced artificial intelligence, with its good and negative consequences for humans (Szocik 2020d), (Szocik and Tachibana 2019) (Szocik and Abylkasymova 2022b).

It is in this sense that I believe our choices today of one path or another lead to far-reaching consequences that are not always recognized. This may be especially true for missions much more ambitious than those being considered today—planets outside the solar system, for example. With respect to such a distant future, we can say that not without responsibility will be those who block the implementation of scenarios requiring the application of gene editing and thereby prevent us from maintaining a life worth living, or bring us to a state where our lives are worse than they would have been were it not for their blocking actions.

In the book I make many assumptions. I assume that humanity will intensify its activities in a sequence of events leading to the exploitation and exploration of space. I thus assume that space will be an essential part of human life and activity in the future, perhaps even something routine. I assume that there will be strong rationales in favor of human long-term deep-space missions even if, in light of today's assumptions, our knowledge of space, and the countermeasures we possess, we may not really see a strong rationale in favor of human space exploration, and lean toward robotic expeditions in light of the risks (Abney 2017, 355–356). I also assume that biomedical enhancement will be necessary, or at least recommended, for long-term space missions, which will be due to either the lack or ineffectiveness of conventional countermeasures. I also assume that biomedical technologies for modifying humans, and gene editing in particular, will be developed with increasing intensity, will be applied, and will be, at least in theory, ready for application.

These assumptions about our future in space and the development of human enhancement technology are key here, because they require recognition that long-term and deep-space travels will be attractive to humans[2] (for scientific reasons, financial reasons, or even as a refuge from disaster on Earth) and that advances in medicine and biotechnology will be so great as to make possible many effects that today are only in the testing, conceptual stage, or simply impossible.

This is a book about the future, space, and bioethics. I understand bioethics as ethical considerations of biomedical procedures (DeGrazia and Millum 2021, 1). In the book I narrow it down to human enhancement, mainly gene editing, applied to space missions. But there is no doubt that these considerations are as useful as possible for a terrestrial context, they can show some unseen effects of various applications, or show that some boundaries are difficult to define between the necessary and the optional, the useful and the seemingly extravagant.

In summary, consideration of bioethical challenges during future long-term space missions is embedded in futures studies. Exploration and perhaps even settlement in space are an integral part of humanity's future.

Overview of Issues Discussed in Space Philosophy and Space Ethics

A need for space bioethics is seen the best when we take a look at the current state of art in space philosophy and space ethics.[3] We can find that bioethical issues are still not considered, or are only minimally considered in space ethics. The current space ethics is dominated by environmental ethics. Space philosophy is, in practice, the same as space ethics (Green 2020). And space ethics is, in large part, the same as environmental ethics. Consequently, when someone is talking about space philosophy, she is, in fact, talking about space ethics, and she has very big chances that she is talking particularly about some issues in environmental ethics related to the context of space missions, human or uncrewed.

The issues most widely discussed in space philosophy and space ethics are those of environmental ethics and extraterrestrial intelligence (ETI) encounter scenarios.

Environmental ethics discusses issues such as, but not limited to astrobiology[4] and extraterrestrial life (ETL), microbial life in space (Chon-Torres 2020), and the distinction between intrinsic value and instrumental value (Peters 2019b). Environmental space ethics also discusses issues such as the need for environmental assessment (Kramer 2020), the value of astronomical objects regardless of possible existence of extraterrestrial life, geoethics, and the concept of integrity (Milligan 2015a), planetary protection (forward contamination and backward contamination), and the problem of space debris and lunar environmental ethics (Norman and Reiss 2021). Christopher McKay defends the intrinsic value of extraterrestrial life (McKay 1990). Charles S. Cockell (2016b) defends the intrinsic value of space microbes on the base of their interest. Kelly Smith (2016) rejects the concept of intrinsic value of extraterrestrial microbial life.

Another large group of issues widely discussed in space philosophy and space ethics, namely ETI encounter scenarios, covers issues such as Search for Extraterrestrial Intelligence (SETI), Messaging Extraterrestrial Intelligence (METI), "should we message ETI?," and considering scenarios and protocols in the event that ETI turns out to be peaceful or hostile to us (Rappaport et al. 2021), (Szocik 2022). Smith and John W. Traphagan propose the concept of passive protocols and inaction (Smith and Traphagan 2020).

One important issue is the rationale for space missions.). Since there are many issues being discussed here, it is best to divide this topic into short-term and long-term perspective issues. Under the former, topics such as the following are discussed, among others: resource utilization and space mining,[5] sustainability and resource depletion on Earth, the issue of spin-offs, and the value of scientific space exploration (Schwartz 2020), (Munévar 2022).

The long-term perspective addresses such issues as, among others, whether we should colonize Mars (Zubrin 1996), (Abney and Lin 2015), (Milligan 2011) and why we should not colonize Mars (Billings 2017, 2019), (Futures 2019), the concept of space refuge, our innate sense and urge to explore space (Zubrin 1996), or some alternative approaches to space colonization. The latter include the negative impact of Abrahamic religions' and Western worldview that consumption is good and unavoidable (Traphagan 2016), (Schwartz et al. 2021); the view of other religious traditions on the issue of expansion in space, such as Buddhism, among others (Impey 2021); and the view of a planet as disposable.

In addition to outlining these three major topic areas, it is worth mentioning that there are some issues among them that have lost their former momentum, such as the question of terraforming Mars (Beech et al. 2021), (Szocik 2021c) or the debate over whether life is better than no life in space, and whether we should therefore seed life on Mars (McKay 1990), (Milligan 2016).

Finally, there are issues in space philosophy that are discussed by individual authors. These issues include the question of astronaut enhancement (Abney and Lin 2015), the issue of genetic liberty and deontogenic ethics in a 500-year space colonization program using genome editing (Mason 2021), the value of scientific exploration of space (Schwartz 2020), (Munévar 2023), and finally the issue of human enhancement for future space missions.[6]

In recent years, several books have been published, both single-authored monographs and anthologies of texts treating space philosophy and space ethics in general, and analyzing in a multifaceted way some selected aspect of space ethics: (Arnould 2011), (Cockell 2015a, 2015b, 2016a), (Milligan 2015b), (Schwartz and Milligan 2016), (Szocik 2019, 2020b), (Rappaport and Szocik 2021), (Green 2021).

In sum, space bioethics, particularly as understood in the manner presented in this book, is absent from the current philosophy and ethics of space exploration.

What Is Bioethics, and What Is Space Bioethics?

It is worth clarifying some terminological issues even if they seem fairly obvious. I was somewhat surprised to find that both in some reviews of my work and in some papers, the term "space bioethics" is used to describe issues discussed within environmental space ethics. This is an example of misuse of terms, in fact a serious confusion of terms.

First, bioethics is not the same as ethics. For some, bioethics may be regarded as an ethical discipline, as an example of applied ethics. Nevertheless, bioethics as such is a separate discipline, moreover, it is not an ethical or, more broadly, a philosophical discipline, however, it may be associated in public space with philosophy and ethics sooner than, for example, with law or medicine, especially when bioethics is dealt with by philosophers who apply moral theories to it. Nevertheless, bioethics is an autonomous science of interdisciplinary character, and philosophy and ethics are, but not necessarily, two of the disciplines included in bioethics. Bioethics understood in this way is not present in the hitherto widely understood philosophy and ethics of space missions. Space philosophy and space ethics cover a number of issues, but other than bioethics, there is no mention of ethical issues related to human biology, which is what bioethics deals with.[7]

Second, space bioethics has nothing to do with astrobiology, although one may encounter misuses of the term bioethics in the context of space missions. Astrobiology is the science of life in space for all possible beings beyond human life and other terrestrial organisms. The scope of interest of astrobiology generally does not include humans as a direct object of astrobiological study. But astrobiology is not an ethical or philosophical discipline. Thus, equating space bioethics with astrobiology contains a double error. Space bioethics covers ethical issues related to human biology in space, whereas astrobiology deals with the origin, origins, development, and altogether the possible existence of life in space. But this life in space that interests astrobiology does not concern terrestrial life as the main and direct object of research. That is what biology deals with.

Another example of the use of terms that can be misleading is the term "astrobioethics." This term is used, among others, in publications by Octavio A. Chon-Torres (2018, 2021b, 2021c). It should be emphasized that there is nothing inappropriate about the use of this term as long as everyone is aware of the meaning of both the term astrobioethics and the term space bioethics. Chon-Torres is well aware of the meaning of the term "astrobioethics," defining it as follows:

> Astrobioethics is the ethical branch that is in charge of studying and analysing the moral implications of astrobiology, such as the evaluation of what to do in the face of a possible contamination that puts at risk any possible form of Martian life. For this reason, we understand as "astrobiological interest" any aspect that involves interacting with any potential life form or conditions that allow us to understand the origin of life in the universe. (Chon-Torres 2021c)

Thus, astrobioethics for Chon-Torres is essentially the same as environmental space ethics, concerned with the status of extraterrestrial life, the ethical challenges that may arise when humans encounter/interact with extraterrestrial life, especially when threatened by humans. These issues are certainly not among those traditionally understood under the name of bioethics, nor are they among the issues I discuss in this book as space bioethics.

To defend Chon-Torres's use of the term, one might add that the prefix "astro" gives the term "bioethics" a new meaning. In an etymological sense, this might be justified by reference to "bio" and "ethics" treated separately. Nevertheless, the term "astroethics" seems to be a more appropriate term here, free from misunderstandings about the application of astrobioethics.[8] Moreover, even if we agree that etymologically "astrobioethics" does indeed mean ethics concerning life in the cosmos—after all, the word says nothing about the fact that "bio" must apply only to humans—the term conflicts with the well-established use of the term "bioethics" applied to humans since the 1960s.

Bioethics grew out of the development of medical procedures in the 1960s that made it possible to prolong the lives of previously lifeless patients, as well as reproductive technologies that went beyond traditional sexual reproduction between a man and a woman (Kuhse and Singer 2009, 3). This has given rise to many ethical issues—specifically bioethics, physician and patient rights, autonomy, accountability, informed consent, and many others. This is what bioethics deals with, and space bioethics is a specific extension and adaptation for the space mission context.

Admittedly, we do not know whether some life form somewhere in space beyond Earth is not facing the same problems and making analogous bioethical considerations. But as long as we do not know for sure, let us agree that we use the term "space bioethics" only to describe ethical issues concerning human biology in space, and save "astrobiology," "astrobiology ethics," and finally "astrobiology bioethics," for describing bioethical considerations undertaken by extraterrestrial intelligence (EI), but only (1) if such an intelligence exists anywhere, (2) if we ever learn about it (after the fulfillment of the first condition), and finally (3) if this EI will carry out such bioethical considerations (after the fulfillment of conditions 1 and 2).

In summary, bioethics addresses the ethical challenges generated by biomedical technologies, while space bioethics brings these considerations to the environment of space mission.

Overview of Chapter Topics

In chapter 2, I briefly outline the specifics of the space environment and the health risks to astronauts. This overview provides justification for my main premise—the hypothetical necessity of human enhancement for future long-term deep-space missions. The chapter also outlines the general methodological framework of space bioethics, indicating that it should use mixed methods and approaches, and be issue-driven rather than theory-driven.

In chapter 3, I discuss the issue of human enhancement. Human enhancement is presented as a rational and pragmatic option that should be seriously considered for future space missions. Perhaps the application of radical human enhancement will be inevitable for an effective human presence in space.

Chapter 4 is devoted to a specific form of human enhancement, namely, embryo selection and GGE. Here I discuss the arguments for and against, although the vast bulk of them pertain to human enhancement per se and are discussed in chapter 3. I also point to perhaps unique applications of these so-called positive selection methods for future space missions.

In chapter 5, I discuss the relationship between the importance of the degree of justification for space missions carried out for scientific, commercial, and to save humanity and the justification for applying human enhancement. I show that such a relationship exists. Thus I emphasize the specificity of space bioethics, in which the context and justification of the application of a biomedical procedure plays a central role.

Chapter 6 deals with the differences between space bioethics and bioethics concerning earthly problems. I point out that there are legitimate similarities between space mission bioethics and bioethics of extreme environments on Earth, such as Arctic expeditions and military ethics. However, I suggest that there are some differences, and I am inclined to conclude that these are differences of degree rather than ontological differences.

In chapter 7, I depart somewhat from the focus promised in the introduction on health-related human enhancements only and discuss the possibility of justifying moral bioenhancement for space missions. I point out that there are many rationales in favor of such an application, but also some ethical nuances that can make this concept a dangerous tool.

Also, chapter 8 maintains the atmosphere of the further future, where I consider the concept of space refuge, but in a specific conceptual context—the philosophy of antinatalism, population ethics, and the ethics of future humans. The main idea of this chapter is this: if the quality of life in a future space colony is too low, we have good reasons for not seeking to save humanity through space colonization. But how do we know if life in space will actually be worse than life on Earth, especially in the situation when the motive for establishing a space colony is a drastic deterioration of living conditions on Earth or even the impossibility of further continuation of existence on Earth (at least as a life worth living)? In this context, I also discuss the possibilities offered by human enhancement for countering antinatalist objections.

The book also includes an appendix on futures studies. The appendix can be read either before or after the book. It is an important supplement to the discussed topics as it shows how space bioethics fits into, and in a sense is a part of, futures studies.

When I talk about space missions, I am referring to future space missions, lasting at least a few years, that will be deployed to Mars or even beyond, possibly to asteroids seen as attractive from a space mining perspective. What is important here is the deliberate adoption of the perspective of space expeditions that will be carried out much further than the missions currently being carried out, that is, beyond the orbit of the Earth and beyond the Moon. Mars is the most obvious and legitimate choice here. However, in this book I will not go into a consideration of the sensibility or validity of such a choice. Nor will I give into such details when I take up the subject of commercial exploitation of space, including Mars, talk about scientific exploration of Mars, or the challenges to human health and life associated with that planet's surface and atmosphere.[9]

In summary, despite the orientation toward considerations of the future, my focus in the book is on what can realistically happen in the future, intentionally eschewing considerations of a lower degree of reality.

2

Human Health Risks in Space and the Methodology of Space Bioethics

Introduction: Long-Duration Human Space Missions

I make the following assumptions in this book: (1) humanity will pursue long-term and deep-space missions; (2) the space environment is and will continue to be too dangerous for humans, at least given conventional countermeasures;[1] (3) various types of biomedical human enhancement will be available, or at least humanity will be working hard on their development and possible applications; (4) the accomplishment of at least some types of long-term space missions will require the application of human enhancement, which at least today is considered ethically controversial.

I argue that planned-in-next-decades human spaceflights to Mars, and in the future beyond, may require application of substantial forms of human enhancement. The context of space missions may be a kind of case study for ethical assessment of the idea of human enhancement which today is discussed in rather abstract context. It is possible that the first enhanced humans will be the first deep-space astronauts. Future human space missions to Mars and beyond may be realized for different research, economic, political, or survival reasons. Since space remains a hazardous environment for humans, space exploration and exploitation requires the development and deployment of effective countermeasures.

The subject of my interest is bioethical challenges only with regard to long-term and deep-space missions, that is, missions lasting at least one year, as well as missions beyond Earth orbit. Thus, the subject of my interest is missions that have never been flown before, because missions lasting more than a year have only been flown in orbit, and those beyond the limits of Earth's orbit, such as missions to the Moon, lasted only a few days. Therefore, the topic of my book equally overlaps with so-called futures studies and is part of what can be called the philosophy and bioethics of the future or reflection on the future of humans in light of new technologies.

Although I will devote chapter 6 to the question to what extent the environment of space missions is ethically unique and to what extent it resembles selected moral environments of the Earth, it is already worth emphasizing that the environment of so-called deep-space is unique in its physical, health, and thus also ethical (although it does not yet follow its bioethical) aspects. These unique features of the deep-space environment are (1) the presence of unknown risks, (2) the lack of proven effectiveness

The Bioethics of Space Exploration. Konrad Szocik, Oxford University Press. © Oxford University Press 2023.
DOI: 10.1093/oso/9780197628478.003.0002

of preventive and protective measures against those risks that it can define, (3) and human isolation (Ball and Evans 2001, 2).

Points 1 and 2 challenge the precautionary principle and, depending on the preferred approach, may suggest either postponement of space missions or pre-emptive application of human enhancement. Point 2, in particular, warrants application, or at least discussion of the possibility and sensibility of application of human enhancement. Finally, point 3 identifies the challenges of isolation and also may argue in favor of human enhancement, which in the context of isolation can be understood in two ways: to minimize the sensation of isolation by applying enhancements, as well as to minimize the health risks, especially of clinical care in space, that result from the inability to return to Earth quickly (or at all) by applying human enhancements. Long-term extra-orbital missions make a qualitative, not just quantitative, difference compared to current missions, and the continuing lack of complete knowledge and certainty is a major challenge here (Ball and Evans 2001, 3).

In summary, the inadequacy of countermeasures and the specificity of long-term deep-space missions open up space for considering human enhancement as a reasonable option.

Space Environment and Health Challenges

Space is not a livable place for humans not only without a life support system, but also without protection against cosmic radiation. Let's look at the example of Mars, considered as the best and most realistic location for a future human habitat in space. The Sun's ultraviolet radiation reaches the surface of Mars unstopped by a very thin atmosphere. Such radiation is deadly to life. In addition, the soil on Mars is oxidizing, which, combined with ultraviolet radiation, destroys organic compounds. Consequently, even the most primitive—and therefore more resilient than forms as complex as human being—life forms could survive on Mars only below the surface, in protected areas such as hydrothermal systems or aquifers (Carr 1996, 170–171). Life forms, in order to survive the threat posed by cosmic radiation on Mars, would have to somehow adapt to these harsh conditions, either through the aforementioned life in subsurface niches or in dormant form. In contrast, the existence of life is plausible insofar as microbes can survive anywhere on Earth where water is available, regardless of chemical or physical conditions, and the same is theoretically possible on Mars (Carr 2006, 272).

The space environment, both in interplanetary travel and on the surfaces of other planets, such as Mars, contains such elements harmful and threatening to humans as different types of cosmic radiation, altered gravity, continuous stay in a closed and small spaces without the possibility of leaving, isolation, and distance from Earth. Humans are simply evolutionarily unsuited to confront such environmental conditions. Consequently, each of these environmental factors exposes astronauts to problems in essentially all areas of health, with cosmic radiation remaining the most dangerous factor.[2]

The Committee on Creating a Vision for Space Medicine during Travel beyond Earth Orbit identifies the following three biggest health challenges in space: space radiation, bone mineral density loss, and behavioral adaptation. As the committee points out, failure to address these three problems will make interplanetary missions impossible. The second challenge is caused by altered gravity and results in changes at the rate of losing 1 percent mineral density per month (Ball and Evans 2001, 3).

In contrast, in the case of difficulties with behavioral adaptation, it can be seen that this risk opens the door to a discussion of the validity of the concept of moral bioenhancement. While I'm not saying that any space agency will ever go for it—in fact, they certainly will not if public ethics remains unchanged, but after all, things could change in the decades to come—I raise this point as an interesting thought experiment that shows how moral bioenhancement might work in a specific, closed environment unlike any other on Earth, where the existence of multiple rationales for and against space missions will be an important element. What if it is precisely the difficulty of behavioral adaptation to the conditions of the space environment that will be raised as the main argument against such a mission, and proponents of biomedical moral enhancement propose it as a solution to the problem?

Similar questions can be raised about the negative impact of altered gravity and cosmic radiation when it becomes apparent that either no alternative and effective countermeasures exist, or their production, theoretically possible, would put many decades or centuries off the realization of long-term space missions (such as the creation of artificial gravity through centrifugal acceleration (NASA 2021) or pose a serious challenge, if not a threat, to the other key factors necessary for space missions (e.g., significantly increasing the thickness of spacecraft walls to more effectively protect against cosmic radiation, which would probably preclude flight until new generation engines and a new source of propulsion are invented (Slaba et al. 2017)).[3]

In conclusion, it can be said that the space environment is characterized by the presence of constant, very dangerous factors as space radiation and altered gravity, which justify the option of human enhancement for future long-term missions of astronauts.

What We Can Modify, and Why

The greatest challenge to human biology in space is cosmic radiation (Crusan et al. 2018), (Szocik and Braddock 2019), (Mason 2021, 114). One option for modifying humans to increase their resistance to cosmic radiation is genetic modification. Modification could involve implementing p53 genes from elephants and/or Dsup genes from tardigrades, which have high radiation resistance (Mason 2021, 108).[4]

I believe that the scope of modifications applied for the purpose of space exploration should be limited only to those modifications that are actually required for the purpose of saving humans from health impairments.[5] Protection from the negative effects of altered gravity or cosmic radiation is the main point of reference and justification for human enhancements here. In one of my papers, I proposed the notion of technological duty, meaning that whether or not the application of human enhancement for space missions is consistent with our moral intuitions, we have no other choice if certain types of long-term space missions are to be pursued (Szocik and Wójtowicz 2019).

More controversial and unclear remains the modification of humans to enhance their efficiency. Modifications can theoretically be applied for this purpose as well, but only under the condition of proving that they are necessary, and that the expected decrease in efficiency will be so great that it may not only harm the success of the mission, but even endanger the health and lives of astronauts by triggering a sequence of events leading to unfortunate accidents. This is why I don't even seriously consider concepts such as, but not limited to, cyborg, cyborgization and "cyborg enhancement technologies" (Barfield and Williams 2017) because we lack proof of dangerous inefficiency.

Certainly, an astronaut or a future space settler who is modified even at the GGE level to protect herself from the harmful space environment will not be a cyborg. But anyone equipped with neuroprostheses or brain-computer interfaces could be considered a cyborg. However, I fail to see the validity of using them for even long-duration and remote space missions, assuming that technologies external to the human body will suffice for computational and other cognitive functions. Imagine a bank employee who, instead of using a calculator as well as the special programs she might need, is fitted with an invasive neuroimplant or interface. I assume the legitimacy of this analogy to the space mission environment. I consider it appropriate and reasonable to apply neural enhancement that has therapeutic justification. In contrast, I find no justification for the superenhancements eagerly discussed by philosophers, such as those aimed at, for example, the ability to detect distances to objects or approaching objects, or the ability to control a wheelchair or a group of robots with neural signals (Warwick 2020). In the penultimate case, the exception may be health reasons that preclude the possibility of traditional manual control and make control by neural signals the only therapeutic possibility.

The moral status of cognitive enhancement for space exploration may leave a margin of ambiguity. This is primarily related to what we consider cognitive enhancement to be, how much of it will be truly necessary (impossible to replace by external devices or training), and how much of it will be redundant (easily replaced or supplemented by external devices and systems).

Cognitive enhancement in its basic sense, meaning the modification of currently possessed cognitive functions includes, but is not limited to "decision-making, reasonability, memory, judgement, situational awareness, attention span, and complex problem solving" (Latheef and Henschke 2020). Thus, so understood, it has morally, pragmatically, and socially desirable goals. Moreover, increasing the effectiveness of these capacities not only serves them or the particular tasks in the performance of which these functions are activated. They are seen as tools and means leading to an increase in the scope of human autonomy as well as the degree of moral decision-making. Both brain–computer interfaces and noninvasive brain stimulation have applications in increasing the computational capacity of the human brain or enhancing the ability to divide attention, especially in task overload and dynamic situations (Latheef and Henschke 2020). In this respect, the dynamics and specifics of the battlefield resemble the environment of space missions, as evidenced by the overload of astronauts on the International Space Station. It is one thing, however, to pharmacologically enhance concentration or attention, but quite another—if it were possible from the standpoint of medical technology—to genetically modify or invasively implant to enhance cognition, whatever that might mean.

Still other possibilities than human enhancement are offered by synthetic biology. The rationale for synthetic biology is quite similar to the rationale for human enhancement. Tim Lewens believes that synthetic biology is not a qualitatively new phenomenon from the point of view of bioethical issues, but "it involves a further step towards one end of the 'design continuum': it involves an effort to apply rational design principles to the organic world" (Lewens 2015, 78). Thus understood, synthetic biology realizes Mason's idea of "genetic liberty" (Mason 2021). It may also be necessary to save humanity from extinction if it turns out that not only human enhancement but even synthetic biology is required to enable humans to settle in space. Synthetic biology may also be required to enable humans to become a multiplanetary species in a variant that is noninvasive to the population remaining on Earth. This option would involve sending instructions to assemble humans from scratch in a new space environment (Anomaly 2020, 72). It is worth keeping this option in mind as a technological possibility to realize the concept of saving the existence of the human species in the distant future, however specifically understood.[6]

In conclusion, human enhancement is understood as a necessary tool, not an extravagant or trivial one, considered as a necessary means to safely and effectively accomplish space missions.

Methodology of Space Bioethics

The methodology of space bioethics deserves a separate book to be written. Here, out of necessity, I must limit myself only to outlining the general framework that guides my methodological approach to bioethical challenges in future long-term and deep-space missions. Just as bioethics in general emerged as a result of philosophical and ethical reflection concerning the legitimacy and possible controversies in the application of new medical technologies (Callahan 2004), so it can be assumed that the bioethics of space missions will "take off" somewhat unexpectedly with the new stage of human development—the entry of humanity into deep space and long-term, perhaps permanent sojourn in a deep-space. This new technical state of affairs—for it is evidently caused by advances in space technology—while not in itself constituting any novel medical procedure of the kinds considered in the birth of "classical" bioethics in the 1960s, will certainly constitute a new inspiration for bioethical reflections.

In my understanding of space bioethics, I refer to John McMillan's conception (which in turn works on Margaret Battin's theory), in which bioethics is a pragmatic and reactive discipline, focused on developing and elaborating moral rationales for relevant and pressing issues (McMillan 2018, 11–12).[7] Also, following McMillan and other bioethicists, I believe that no ethical theory should determine bioethical considerations (McMillan 2018, 35). Thus, space bioethics, like bioethics in general, should be issue-driven, not theory-driven. I also refer to the understanding of bioethics proposed by Ruth F. Chadwick and Udo Schüklenk who point out that bioethical analyses need not be and usually are not as deep and complex as metaethical considerations. But the task of bioethical deliberation is to take some perspective, such as a chosen normative concept, apply it to the biomedical cases being analyzed, and then see where the argumentation takes us (Chadwick and Schüklenk 2021).[8]

Another important criterion is "engagement with experience" and "engagement with reality" (McMillan 2018). McMillan seems to be critical of thought experiments on bioethical issues, such as the famous violinist experiment in Thomson's 1971 article. As McMillan himself notes, the essence of philosophical thought experiments is to draw attention to the correctness of reasoning and the clarity of concepts, not to describe real-life situations. However, this does not change the fact that they are far from reality and sometimes it can be difficult to embody the situation discussed in a given experiment by means of abstract analogy.

Space bioethics is specific, however, because of the object of its study, namely potential challenges in future space missions. Thus, it is important to distinguish between what McMillan says about the application of thought experiments to currently happening biomedical procedures and the situation in which a biomedical procedure is considered *in abstracto*, as not yet happening in reality. Since my goal is to outline the potential effects of various alternative future scenarios and to show that each has certain bioethical implications, I use some thought experiments or hypothetical examples. However, this is due to the specificity of the subject matter. In any case, the constant reference to experience is due to the fact that I am considering realistic alternatives, however far in the future they may be.[9]

For these and other reasons bioethics must be empirical in nature. Being theory driven unnecessarily brings bioethics closer to such "unreal" branches of philosophy as metaphysics or epistemology, which can function perfectly well without reference to empirical knowledge. Bioethics, though, must be "practically normative" (McMillan 2018, 94). This empirical context of bioethics is not only connected with its interdisciplinarity, but—in the context of space bioethics—it also results from the interdisciplinarity of space philosophy and space ethics (Schwartz and Milligan 2016; Schwartz 2018, 2021).

I assume that the most optimal methodological perspective for space bioethics is the issue-driven orientation, which consequently leads me here not to consider arguments inherent in ethical theories based on a few specific principles (theory-driven approach and principle-based ethical theories).[10] I take a perspective that analyzes a particular situation, place, and time, as well as the context of the people involved in a given ethical situation, in the way considered by care, or feminist theories (Veatch and Guidry- Grimes 2020), but I also examine whether the classic four principles of bioethics apply (whether they can, should, or must) in selected contexts of future space missions.[11]

My approach to moral norms and ethical and bioethical theories is close to what Robert M. Berry calls "fractious problems," which are highly complex, in principle almost impossible to solve unambiguously with a chosen theory, generated by advances in science faster than our moral intuitions. Berry cites the views of Alasdair MacIntyre and Thomas Kuhn on a kind of insolvability of issues in contemporary moral debate and values. He cites Macintyre talking about the incommensurability of systems. This may lead to a situation in which different ethical theories offer different answers to the question of acceptability of the procedure or behavior under analysis. But, as Kuhn quoted by Berry says, there is a set of values to which the conflicting sides of the scientific debate refer (Berry 2007, 2–3).

My approach to ethical norms and theories is closest to some version of particularism. Particularism is often heavily criticized (plea of skepticism), primarily because it leads to an infinite regress and does not really provide a definitive answer to the question of the basis for a given norm (Tännsjö 1998, 12). However, as I point out in the book, there are good reasons for recognizing some contextual value of particularism and specification in relation to the space mission environment.

Preference for particularism and an issue-driven approach does not mean the exclusion of any principles. We usually agree that the moral duty is to protect human life or not to cause suffering, unless justified by necessary defense or medical procedures designed to cure the patient. In the book, I often refer to the four principles of autonomy, beneficence, nonmaleficence, and justice proposed by Tom L. Beauchamp and James F. Childress (Beauchamp and Childress 2013).

It is not difficult to imagine a situation involving space exploration in which, as in many bioethical cases on Earth, we will have to deal with the conflict between principles, rules, or duties. If at least two duties appear to be prima facie duties, how to identify duty proper in the context of space exploration?[12] Acceptance of the non-consequentialist approach can have negative consequences for mission success. The duty-based approach will lead to a situation where, in case of conflict with the consequence-maximizing approach, certain duties will be considered more important than the desire to maximize desirable consequences (Veatch and Guidry-Grimes 2020, 73). The mentioned four principles involve potentially conflicting ethical approaches, consequentialist and deontological. For example, the principle of autonomy as such is in conflict with principles of beneficence and nonmaleficence (Chadwick and Schüklenk 2021, 30–32).[13]

It is accepted that the main task of bioethics is to warn about the potential risks of applying new biomedical technologies. But it is useful to change this perspective from one that denies to one that affirms, and to see bioethics as an advocate for new technologies that have potential benefits. In this approach, what is risky is not the application of the technology, but the lost benefits of its application (the so-called opportunity-cost) (Ravitsky 2021). As I show in the book, space bioethics should first of all be open to the possibilities guaranteed by space missions, as well as to the radical human enhancement that even makes them possible. This is the starting point. At a later stage, however, space bioethics should carefully evaluate potential threats posed either by space missions or by human enhancements applied for their purposes.[14]

In sum, the methodology of space bioethics should focus on a case-by-case analysis considered in particular circumstances, without prejudging a preference for any normative theory, principles or rules.

Mixed Methods and Issue-Driven Approaches to Space Bioethics

Many issues concerning future long-term space missions today are still unresolved. For we do not know what type of missions we will pursue, if any, and we also do not know whether we will produce conventional countermeasures that would allow us to forego applications of what are today considered controversial biomedical human enhancements.[15] Finally, we do not know whether we will have all the desired and effective biomedical enhancements in a future in which such missions are technically feasible, nor whether they will continue to be considered controversial. Under such conditions, the optimal approach is to apply mixed methods to the space bioethics, a syncretic approach that will not favor one chosen perspective.

Let us begin with what bioethics has encountered with respect to biomedical controversies on Earth. Many procedures that are treated routinely today, and are often recommended and desired by both patients and physicians, that is, organ transplants, in vitro fertilization, withdrawal of life support, and many others, were no less controversial in the beginning than human enhancement through genome editing is today. Despite this controversy, over time these procedures have become accepted. This shows that what drives health policy and public health law is consequentialist thinking. For we look at whether we will get better consequences in the world by banning transplants or in vitro fertilization or by allowing them.

Deontological thinking is also interwoven here, at least as long as we have to recognize as a moral obligation the duty to protect human life. However, first, these duties can be derived from the consequentialist principle of beneficence, and second, even if the principle of the protection of life in the context of healthcare were indeed duty-based principles, the deontological approach is a major opponent to new biomedical technologies. For often, as I will show in the next chapter, bioconservative opposition to human enhancement is derived from principles such as the protection of dignity, the inviolability of the natural, autonomy, or justice. I believe that deontologism is valuable as an expression of prudential skepticism, but at some stage, with both the safety of the biomedical procedure in question and agreement about its guaranteed benefits, it should be rejected as an approach that impedes the development and advancement of quality-of-life ethics.

In such a way, we can say that in this syncretic, methodologically open approach to space bioethics, if we refer to the two main theories of right action, namely consequentialism (consequence maximizing principles) and deontologism (duty-based principles), space bioethics favors consequentialist rather than deontological thinking. Nonetheless, sometimes some good consequences should not be allowed when they would require the suspension of certain principles, such as the protection of human life or autonomy. I do not prohibit or make mandatory any human enhancement procedure, but I do draw attention to the consequences and to whether these critical principles may be violated in individual cases.[16]

As an example, let us look at the now controversial concept of GGE. I am not prohibiting or demanding the application of this procedure to space missions. Instead, I analyze the consequences of two scenarios, one that forbids this application and one that permits it, and draw conclusions about whether GGE might be morally permissible, or perhaps even desirable, under given conditions. In contrast, I do not believe that any human enhancement has the status of a moral obligation, because I concede that there can always be at least subjective objections of some individuals and these should always be taken into account. Recommendations can be derived from my approach to space bioethics, but certainly not moral obligations. A person must always have an alternative and cannot be forced into a given action even if it has only good consequences.

On the other hand, where, in my opinion, the application of a given biomedical procedure will lead to a scenario with better consequences than the alternative of forbidding the procedure, I recommend such human enhancement in a given context. In this sense, space bioethics can be said to have a consequentialist tinge, but even if it is a statistically dominant way of analyzing biomedical issues, it is decidedly not absolute and sometimes gives way to, for example, deontic ethics. Indeed, space bioethics is context-dependent.

Space bioethics can also make good use of contractualism. Rawlsian contractualism assumes an initial lack of knowledge about the details of a given moral situation, and most importantly, a lack of knowledge about our position and standing in the particular situation being considered as a moral dilemma. The concept of the "veil of ignorance" is the key term here, and it expresses the situation in which we must determine what is morally right without knowing whether we would be the beneficiaries of a given ethical decision (Rawls 1999). In practice, contractualism often remains aligned with utilitarianism. Both the proponents of the one[17] and the other theory usually emphasize the necessity of making such moral decisions that favor obtaining the greatest possible good (Huang et al. 2019).

Principlism is similarly useful as an approach that gives a general framework. At least the first versions of principlism were based on the deductive method and involved the application of abstract principles. It was a top-down method, often criticized, which should not be used by space bioethics. But it is difficult to move entirely without any principles in trying to determine the moral status of particular bio-medical modifications. As Allen Buchanan and colleagues rightly point out, simply relying on the judgment and morality of individuals considered virtuous is not suffi-cient in public life to assess the quality of institutions. Reference to principles is still needed (Buchanan et al. 2000, 378).

I don't refer to virtue ethics in general, because I don't know whether a person who prohibits GGE or who accepts it is more akin to the ideal of the virtuous sage, or whether it is precisely that person who recommends that moral decisions be context-dependent. In contrast, good candidates for a moral principle that seems to be ac-cepted universally in space bioethics specifically understood in the context of space exploration are the principles of respect for human dignity[18] (Pullman 1999), protec-tion of life and autonomy (see chapter 4 in this book on GGE and embryo selection, and chapter 8 on the concept of protecting the existence of our species at all costs), and the principle of well-being.

Since I mentioned that space bioethics must be case-driven, associations with case studies may come naturally. Casuistry, along with particularism, are arguably the two most criticized methods in bioethics today. In the case of space bioethics, at least as of today, we have no paradigmatic case. We do have the case of the infamous GGE ex-periment conducted by He Jiankui, but it is difficult to treat it as a paradigmatic case because of its fundamental deficiencies in concern for safety, efficacy, and transpar-ency, and many others (Krimsky 2019). On the other side, it may be a good example of bottom up approach when we try to derive norms or principles from a case study.

On the other hand, the experiment conducted by Jiankui can be a kind of test case for the evolution of people's future behavior and perhaps even the development of public policies. However, looking at the reactions of even hitherto liberal bioethicists (see, for instance, (Caplan 2019)), it can be said that the Jiankui experiment has squan-dered any chance of progress in legalizing GGE (Allyse et al. 2019), (Kleiderman and Ogbogu 2019), (Greely 2021).[19] If it turns out that the three children born as a result of his experiments do not have any complications in the following years of life, this fact may encourage more people to apply CRISPR to modify germ cells in order to eliminate certain diseases or reduce the risk of contracting some of them even if the embryos remain healthy. This may especially encourage those who already benefit from assisted reproduction (Kirksey 2020, 245).

Nor is such a paradigmatic case the NASA Twins Study (Garrett-Bakelman et al. 2019), where, although Scott Kelly stayed on the International Space Station for 340 days, which meets the criteria for a long-term mission, it was "merely" Earth orbit, not a deep-space mission.

Casuistry in bioethics offers so-called the bottom-up approach, in contrast to the top-down methods of applied ethics (Arras 2017, 46). Its primary method is rea-soning through analogies and comparing different cases. John D. Arras lists as key steps in casuistry methodology defining a paradigmatic case within an existing tax-onomy and then applying analogies to existing similar situations (Arras 2009, 118–119). I think this approach has the potential to be used by space bioethics, especially when we want to evaluate the moral and legal legitimacy of biomedical procedures even such as GGE for preventive purposes. A case-by-case approach allows the rejection of cases that pose a risk but does not reject the chance of eliminating a serious hereditary condition (Gyngell 2017). Nevertheless, it is also worth remembering that few cases known even from the ethics of extreme environments on Earth are comparable to the situation in space.

Useful approaches to space bioethics are care ethics and feminist bioethics (Szocik 2020c, 2021b). Three basic features of care ethics include the priority of the specific situation over moral norms, the primacy of care among other virtues, and the predominance of virtue ethics over action ethics (Veatch and Guidry-Grimes 2020, 100). What I particularly use in the book is the first of these characteristics. Rita C. Manning mentions as a key feature of care ethics the so-called moral attention which is a form of the aforementioned attaching importance to a specific situation. Moral attention prompts us to learn about each ethical situation in detail. The requirement here is a sympathetic understanding which means a form of imagining the position of the other, feeling into their states, needs, and desires. Manning rightly calls this attitude maternalism as opposed to paternalism, which grows out of the position of "the doctor knows best" and does not necessarily have to reckon with the opinions and states of the patient. Relationship awareness is quite literally the realization that each of us is in relationships of various kinds with others that result in mutual obligation and concern. Harmony and accommodation point to the need to seek a reasonable and fair balance in our demanding and caring actions (Manning 2009, 106–107).

Feminist philosophy is almost entirely absent from space philosophy, which is male-dominated and dominated by topics stereotypically considered masculine, such as colonization, conquest, the final frontier. The use of feminist space bioethics is important in making space bioethics more balanced and less male-centered. Feminist ethics offers a good context for space bioethics because it emphasizes the importance of social context and the role of relationships (Gilligan 1982; Walker 1998). I believe that the main argument for the usefulness of the feminist approach is that it assumes situationism and particularism. As such, it gives no preference to any normative approach, and recognizes no rule for an absolutely determinative ethical analysis (Veatch and Guidry-Grimes 2020, 94).

Finally, space bioethics is to some extent speculative bioethics. Empirical methods in space bioethics, however preferred, are not always possible. Therefore, speculative bioethics remains an inevitable complement.[20] Speculative bioethics is understood as considering future possible scenarios, showing the potential long-term consequences of various ethical choices, and finally using thought experiments (McMillan 2018, 128).

Basically, I take the position that our moral intuitions are valuable and that as a default position we should trust them rather than reject them. But I object to taking this attitude of trusting moral intuitions in a completely uncritical way. One of the problems with intuitions is that they often express approval of a-rational or irrational attitudes, when even a little rational reflection would lead to a different conclusion.

Intuitions can also be linked to biases and subject to change depending on place and time (Benatar 2006, 203). Intuitions, on the other hand, are an essential part of common morality.

One may wonder whether the prohibition of GGE is the result of certain moral intuitions about human dignity, for example, or perhaps the result of paternalism. Paternalism can be manifested by greater or lesser bans on genetic modification technologies or by public institutions blocking in various ways access to information about DNA and the possibilities of modifying it. The public should be informed and make informed decisions (Plomin 2018, 184). The lack of medical and biological and especially genetic education leads to a sense of fear and demonization of genetic technologies, especially since they describe complex processes that only a small part of the population understands (Clark and Pazdernik 2009, 667–668).

Misinformation during the Covid-19 pandemic is a prime example of a lack of knowledge coupled with an irrational following of conspiracy theories and a basically incomprehensible reluctance to accept scientific sources. The problem of misinformation and popular education thus brings into question the issue of paternalism. Should we, in the name of freedom and autonomy, condemn attempts by public institutions to interfere where decisions are made solely on the basis of science? But should we be completely consistent and distance ourselves from every attempt by public institutions to interfere in medical, including biomedical, issues, even when determined by purely scientific reasons? This issue becomes even more problematic in the context of future missions, where there are more unknowns and more at stake (for example, the survival of the species) than on Earth.

How to combine such different views, where one accepts casuistry and care theory on the one hand and deontic ethics on the other? A good tool is the method of reflective equilibrium. This method, originally proposed by Rawls (1999, 18–19, 42–45), has arguably the greatest degree of inclusiveness and comprehensiveness among methods in bioethics. These advantages are at the same time its disadvantages, because in principle one can indefinitely refer to numerous theories, norms, principles, and even a broader social and political context when analyzing particular bioethical issues. However, this theory can be treated as a certain ideal to which one should aspire even if it is not possible—or if it is simply too burdensome and time-consuming in practice—to actually refer to and confront all theories, norms, and principles (Arras 2017, 185–186). In this method, beliefs and moral principles are not derived from a single theory or norm, but their coherence with as many other principles and beliefs as possible is examined.[21] The method of collective reflective equilibrium is an interesting extension of reflective equilibrium and complements it by adding, above all, lay public and citizens to the group of deliberators, thus expanding the elite group of professional ethical theorists that was the starting point for traditional reflective equilibrium (Savulescu et al. 2021).

I also argue that space bioethics engages issues that perhaps more often than bioethical issues around human enhancement on Earth imply the conflict between the rights and welfare of the individual and the rights and welfare of the whole. The conflation of these two perspectives on Earth is more commonly seen in relation to public health ethics. Awareness of the inevitability of the conflict comes automatically when we turn our attention to what is at the heart of current bioethical considerations, and what is at the heart of public health ethics.

Bioethics first of all promotes the principles of autonomy and respect for the individual, while public health ethics—by definition—upholds the health of the population (Blackshaw and Rodgern 2021). It is in this arena that conflicts arise in areas such as recommending or even in some situations compelling acceptance of vaccination (for the good of the public) versus the right to make decisions about one's health and life based on the principle of autonomy and respect for the person. To a certain extent, public health ethics justifies and is based on paternalism[22] (although they are not the same thing, but the paternalistic concept assuming that the doctor knows best is related to the top-down prescribed concern for the good of the whole at the expense of the opposition of individuals) from which bioethics departed rather quickly, in the second half of the twentieth century.

I do not use the term "bioethics" in the sense understood for clinical applications, in the way we talk about bioethics in connection with an institutional review board (IRB) assessing primarily the fulfillment of the criterion of informed consent or ensuring the privacy of medical data. These issues are important to me and I mention them too, but what interests me is rather the search for justification of human enhancement applied for space missions, the search for differences between Earth ethical environment and space mission environment, with the same, philosophical, metatheoretical approach more than clinical in the spirit of IRBs. I believe that issues such as the astronaut's informed consent and the privacy of their medical data become trivial when confronted with an issue such as the concept of radical human enhancement for space missions, incorporating the GGE, embryo selection, and moral bioenhancement concepts discussed in this book.[23]

There may be many approaches to space bioethics, and probably not everyone interested in this topic will find my approach adequate. Nevertheless, I give what I hope are convincing arguments for why my approach to space bioethics, though not the only one, is justified. I believe that human enhancement by biomedical means will be a central issue in space bioethics, and we have good reason to already be discussing the application of human enhancement to future space missions because of the implications it may have in the near and distant future.

I am not saying that what is meant by space bioethics today, which is primarily the issue of an astronaut's autonomy and informed consent as to participation in research during space missions, the issue of privacy of their medical data and acceptable risk, are not important ((Robinson 2004), (Wolpe 2005), (Koepsell 2017), (Spector 2020), (Green 2021, 56–63), (Sawin 2021)). On the contrary, those issues are and will remain very important. However, my point of view goes into the future, goes beyond this clinical, applied bioethics, and points to more distant constellations of issues such as plans for long-term and deep-space missions, threats to the space environment, existential challenges on Earth, and the availability of new technologies, including biotechnology and genetic engineering,[24] that could suit humanity for space travel.

In summary, space bioethics as I understand it is not narrowed to any ethical theory (it is not theory-driven). It is a bottom-up (case-driven) approach that can lead to identical results with top-down approaches that come out of applying normative theory to individual cases. I find useful the dual value theory, which recognizes the existence of two fundamental principles such as well-being and respect for rights-holders (DeGrazia and Millum 2021, 56). Already these two principles when applied to space bioethics suggest that space bioethics is an eclectic approach, combining what is typical for consequentialism (well-being, utility, beneficence and nonmaleficence) with what is crucial for deontologism (respect and autonomy).[25]

I consider a weak version of rule-utilitarianism to be an approach that works in many cases discussed in space bioethics, where certain moral principles and rules like the mentioned well-being and respect, but also autonomy, but also many others that we quite commonly accept within common morality, should apply at least as prima facie obligations. Weak form of rule utilitarianism is considered by me to be the preferred approach to space bioethics, but not the absolute and only one. My theory is not rule utilitarianism in the strict sense, because I do not regard utility as a guiding principle.[26] Often the principle of autonomy and respect turns out to be more important. Thus, I believe that various considered moral judgments (moral rules) are justified at the starting point, but are not absolute (Beauchamp and Childress 2013, 405). I agree that there are some moral constraints, but I also think that the number of absolute moral constraints is very limited.[27]

This approach, focused on analyzing a particular ethical situation, guides me throughout the book even when I do not refer to any particular principle, rule, and theory. I believe that this eclectic approach is, at least for space bioethics, practically more effective than theory-driven ethical approaches that always appeal to the same principles of right actions or virtues. I believe that both consequentialism and deontologism taken in an absolute way may either lead to certain abuses and not always take sufficiently into account the well-being of the individual (risk associated with consequentialism) or limit the horizon of possibilities in the name of particular duty-based principles and rules (risk associated with deontologism).

Finally, an argument for an eclectic, issue-driven approach to space bioethics is that it considers future moral problems. We do not know whether the absolute application of the chosen moral rule will be acceptable in all possible future scenarios.[28]

3
Biomedical Human Enhancement

Introduction

By human enhancement I mean primarily gene editing. Gene editing today is permitted only in exceptional situations for clinical, therapeutic, or preventive purposes. It is reasonable to assume that legal acceptance of human enhancement through embryo selection and gene editing will increase in the future. The very fact that there has already been genetic modification of humans, not only of somatic cells but more recently even of embryonic cells, means that the first step has been taken. Consequently, as John H. Evans points out, there has been a kind of normalization of the idea that humans can be genetically modified (Evans 2020, 11).

I will argue that it is our responsibility to increase genomic and gene-editing knowledge and apply it broadly, and that the space mission context is a new area of its potential application that may open the door to broad application on Earth. I believe that what defends the use of human enhancement on Earth justifies all the more the use of human enhancement for space missions. If a given context and purpose justifies the application of human enhancement on Earth, then one can be sure that its application for space missions will be all the more justified. In contrast, the reverse inference is possible, but there are theoretical situations in which it may not be true. This is because a biomodification justified in space or for space missions will not always guarantee justification for an analogous modification on Earth. But much depends on the context.

In summary, human enhancement by gene editing can be an essential part of human life on Earth and in space, with its application to space having a very different status from Earth application in terms of function, purpose, and justification.

Human Enhancement: Conceptual and Definitional Issues

According to one definition, "human enhancement"

> refers to a variety of efforts—some still best treated by science fiction, some well-established in today's societies—that are intended to boost our mental and physical capacities, and the capacities of our children, beyond the normal upper range found in our species. (Lewens 2015)

The Bioethics of Space Exploration. Konrad Szocik, Oxford University Press. © Oxford University Press 2023.
DOI: 10.1093/oso/9780197628478.003.0003

This definition appeals to a certain consensus in the classical understanding of human enhancement that as long as an intervention leads to an outcome beyond what is typical for the species, it is enhancement, whereas if its function is to bring a trait to its standard degree of intensity, it is therapy.

The next definition is more practical for my approach, for it frees us from clinging to the cumbersome notion of the norm and going beyond the norm or, in the case of health-related improvement, matching the norm: "In bioethics the term 'human enhancement' refers to any kind of genetic, biomedical, or pharmaceutical intervention aimed at improving human dispositions, capacities, and well-being, even when there is no pathology to be treated" (Giubilini and Sanyal 2016, 1).

However, my understanding of the term "radical enhancement" differs from other meanings attributed to the same term. According to the definition proposed by Nicholas Agar,

> Radical enhancement involves improving significant human attributes and abilities to levels that greatly exceed what is currently possible for human beings. The radical enhancers who are this book's subjects propose to enhance intellects well beyond that of the genius physicist Albert Einstein, boost athletic prowess to levels that dramatically exceed that of the Jamaican sprint phenomenon Usain Bolt, and extend life spans well beyond the 122 years and 164 days achieved by the French super centenarian Jeanne Calment. (Agar 2010, 1)

When I speak of radical enhancement, I do not mean modifications that aim to achieve a kind of extreme and maximum of human capabilities in a given trait. For the intent of such modification for space missions is not to seek to maximize the performance of a feature for its own sake, and certainly not for trivial purposes. Thus, my understanding of radical enhancement may lead to the modifications described in Agar's definition. The essence of radical enhancement in space is to guarantee optimization, not maximization.

Agar's definition is in some ways useful. Enhancements applied to space missions can indeed significantly increase the performance of a given feature of human nature. They may also involve the application of new functionality not yet possessed by any human. Thus, while the effect of radical enhancements applied for space missions may correspond to the effect of radical enhancement as defined by Agar, what makes them at least prima facie different is the rationale. Rather than using multiple definitions of human enhancement, it suffices to follow John Harris in saying that enhancement means any change for the better (Harris 2007, 36).

I assume that human enhancement is nothing controversial. The burden of proving the moral controversiality or even the impermissibility of human enhancement that stems from this controversiality lies with the critics and opponents of human enhancement, not with its supporters. I point out that the justification for such a position is the fact that continuous human modification, not only natural but also artificial selection, has been occurring for thousands of years. A correlated justification is also one of the main functions of the development of science and technology, namely their applicability in improving human life. Enhancements of various types are ubiquitous (Bostrom and Savulescu 2009, 2–3), and cultural evolution also produces significant lasting effects on the human genome, thus itself a form of radical and far-reaching enhancement (Richerson et al. 2021).

When I speak of enhancement, I mean only those human enhancements that require and involve close integration with the body, where the modification realized directly penetrates the body of the individual being modified (Erler and Müller 2021). This is one of the features that make human enhancement controversial, namely invasiveness.

The ethical position presented in this book can be described as permissive. This means that human enhancement, even the most radical, should be morally acceptable unless some extraordinary circumstances and obstacles, such as evidence of the harmfulness of a particular biomodification, arise to justify its rejection. But such a circumstance that rejects the permissive position presented here is the situation (described at the end of the chapter) in which the consequences in the world desirable from the consequentialist position are not considered desirable, or are considered less important than following certain rules. In conclusion, human enhancement should be understood as improvement.

Means of Enhancement versus Degrees of Enhancement

An interesting approach to human enhancement is offered by Agar. He points out that the criterion of ethical evaluation of a given human enhancement procedure should not be the way in which we achieve a given effect, but the degree of realized change, its radicalness (Agar 2014, 343). In another work, Agar and Felice Marshall add that the criteria for moral evaluation of human enhancement are as follows: "the *means* by which humans are enhanced, the *capacities* that are the targets of enhancement, and the *degrees* to which they are enhanced" (Agar and Marshall 2015, 533–534). In the following subsections, I will consider the relevance of these categories in relation to human enhancements in general, and human enhancements for space exploration in particular.

The Means

The means by which the modification is carried out is the element that is morally relevant to the discussion on the justification for biomedical modification of humans. The primary difference between traditional enhancement and radical enhancement is that with traditional enhancements, the modifications were not applied directly to the human body. Thus, they were not invasive and did not directly react with the human body in any way. Radical enhancement done through biomedical means changes the way modifications are applied. Biomedical enhancement always involves direct invasive interaction with the human body, whether in the form of pharmacological, genetic modification or implants or neuronal interfaces. Thus, it is clear that the manner in which modifications are applied has moral significance and makes the difference between traditional forms of enhancement and radical biomedical enhancement.

I believe that, with respect to human space exploration, the means by which modification is introduced has moral significance, but in a perspective quite opposite to that in which it is considered on Earth. On Earth, the current discussion presents biomedical enhancements as those means that must be specifically justified, usually by therapeutic purposes, and often when alternatives are not available. I believe that in the case of the space environment, the status and position of biomedical enhancements are reversed. They are privileged, they are the default enhancement modality whose omission rather than use requires special justification.

Biomedical enhancements with respect to the space environment are shown to have advantages over traditional enhancements because of the cost, speed of application, effectiveness, and justification of its application. The means of applying enhancements to space missions refer to the moral aspect, but in the sense that rejecting the application of biomedical enhancements can be treated as morally wrong and hold one morally accountable. I disagree with Agar, for whom the criterion for the moral evaluation of enhancement is the degree of change applied, not the means by which the change is applied (Agar 2014, 142).

Agar's concept of moral evaluation based on degree rather than means of modification is not applicable to contemporary bioethical challenges in light of biotechnology. His concept, as he points out, removes the moral distinction between genetic modification and environmental modification. Thus, both genetic modification and environmental modification can be equally morally unjustifiable if both have been overapplied (Agar 2014, 195). But this posing of the problem leads in turn to the new difficulty of determining the limit and ultimate level of development of a trait. What criteria should be adopted to consider that a given degree of a trait is undesirable?

Harris, in discussing the difference between different ways of achieving the same effect, gives the example of the difference between mechanical enhancements, which are generally accepted, and chemical enhancements, which are either unaccepted or raise questions. He points out that the latter lack justification and that resistance to their use has a kind of "yuck factor" (Harris 2007, 20). Indeed, for some reason the aversion to biomedical moral enhancement may be rooted in irrational moral intuitions. However, moral intuitions have in common that they often have nothing to do with rationality and are difficult to justify, as well as historically variable.

In summary, the only element that accounts for the possible controversy of biomedical human enhancement as opposed to conventional enhancements is precisely the nature (means) of the modification, that is, the use of biomedical instead of conventional means.

The Degrees

It is possible to identify some contexts in which degrees may play a role. But even in those situations it plays a role for reasons other than those given by Agar. In Agar's approach, the degree of modified change plays a role when it leads to the production of superchange—change far beyond the capacities inherent in human biology that make humans superhuman or posthuman.

In the space mission environment, the degree of change introduced per se does not play an important role in ethical justification. Instead, it may be relevant in one specific situation, namely readaptation to the terrestrial environment. This is a special and perhaps unique case in space bioethics, where regardless of the possible moral controversy of the means by which enhancement is applied, its degree is relevant, but only if different degrees of introduced modification determine different degrees of readaptability, or perhaps even preclude it. If, however, the modification in question, even if applied to the highest possible degree, makes readaptation to a terrestrial environment possible, this still does not make the degree of modification applied a morally significant situation. Thus, for the degree of modified change to be considered morally relevant, the assumption of irreversibility is also important. But if so, then it is not the degree that matters morally, but the fact of irreversibility.

We can imagine situations in which a given trait is modified to such an extent that it not only makes it impossible to return to Earth but it would limit their other activities in space or make it impossible to move between different environments. However, it is difficult to predict whether adaptation to the environment of, say, Mars would require the application of irreversible point modifications that would in turn make it impossible to apply other point modifications necessary for survival elsewhere in space. In principle, however, the scenario described falls under the problem of readaptation and reversibility. Possible irreversibility precluding readaptation to terrestrial conditions or adaptation to other places in space would limit the right to free movement, freedom of choice of place of residence or place of work, but also freedom of movement for purposes of tourism, for example. Certainly the limits of such freedom will be determined, at least initially, by technical possibilities, the scope of human exploration of space, and the conditions of life on Earth after a possible catastrophe. Nevertheless, an important question arises here about the ethics of the quality of life in space, one element of which is freedom of movement.

With regard to the plea concerning difference in degree, Harris gives the example of the (lack of) moral difference in the use of reading glasses and binoculars. He asks: is the use of binoculars immoral simply because the person using them has abilities beyond what human sight can achieve (Harris 2007, 19–20)?

In summary, the degree of applied modification is morally neutral, with the rare exception of irreversibility precluding readaptation to the terrestrial environment.

The Capacities (Target and Goal)

I do not consider the issue of the target of modification, that is, the modified genes in the case of human enhancement by gene editing, to be a controversial issue. And it is even less controversial in the context of space exploration, where the goals of modification are primarily functions related to health and resistance to harmful environmental conditions.

Radical Human Enhancement and Its Moral Controversy

Radical human enhancement means a modification that is radical precisely in the sense that it is highly invasive, may be irreversible and significantly alters the quality of a normal, healthy individual. It involves the addition of a significantly new ability, or at least a significant quantitative increase in the performance of an already possessed parameter, by means of methods considered controversial. This last element of the definition—noting the use of a method considered controversial—is crucial.

The concept of radical human enhancement is a context-dependent and evolving issue over time. The example of transplantation or the in vitro method of fertilization is a good example of how the degree of controversy attached to a given medical procedure diminishes and, over time, disappears completely, or almost completely. Buchanan argues that the alleged unique controversy of biomedical enhancement is neither because it is enhancement, nor because it involves moral issues, nor because it causes biological or genetic effects. As he points out, all three of these potentially controversial elements are already present in nonbiomedical human enhancement, which humanity has been applying for thousands of years (Buchanan 2011, 24–25). These observations are important for the discussion of the ethical status of biomedical (radical) human enhancement. They seem to simply elude critics of biomedical enhancement.

Some of the changes are truly radical in nature, primarily in terms of increased life expectancy or education, as well as moral and social behavior. As Buchanan and Powell explain, certain behaviors formerly considered immoral have been made acceptable today in many parts of the world. On the other hand, many behaviors that were once acceptable are unacceptable today (Buchanan and Powell 2018).

Another point mentioned by Buchanan is that nonbiomedical enhancements known to humanity have involved moral issues. Buchanan gives the example of the risk of nuclear war, which can be seen as a byproduct of such enhancement as science (Buchanan 2011, 25). Scientific and technological developments, good in themselves, have more negative effects that generate moral issues.

The third of the controversial features mentioned by Buchanan concerns biomedical enhancement causing biological and genetic changes, such as the shuffling of the gene pool through migration (Buchanan 2011b, 24). All these examples go to show that enhancement is something ubiquitous in human life as a social and cultural being.

The aforementioned life extension is a change of population nature, and therefore, given the high degree of change, is a better candidate to be classified as an example of radical human enhancement than gene editing. Perhaps more significantly, non-biomedical enhancements are, as Buchanan points out, irreversible. The charge of irreversibility is one of the main criticisms leveled against biomedical enhancement, particularly gene editing (Buchanan 2011, 40). An example of such in the context of space is the aforementioned situation of readaptation to Earth or adaptation to other locations in space. The moral controversy over the issue of irreversibility is context-dependent. As Karolina Kudlek rightly justifies, human enhancements cannot be carriers of values in themselves. And even if they were, as she adds, it does not follow from their nature and specificity that their value must be negative (Kudlek 2021).

In conclusion, the objections to even the so-called radical forms of human enhancement weaken when it is shown that many of the modifications applied through conventional methods are permanent, irreversible, and profound in nature.

The Welfarist Theory of Human Enhancement

Julian Savulescu and colleagues propose the following welfarist definition of human enhancement: "Any change in the biology or psychology of a person which increases the chances of leading a good life in the relevant set of circumstances" (Savulescu et al. 2011). According to this approach, enhancement is meant to improve well-being. So what is crucial is to define what a good life means. Once this is established, the idea of enhancement itself is hardly controversial. Instead, one can dismiss particular types of enhancement as not serving the good life.

The welfarist approach to human enhancement has another interesting consequence: namely, it seems to completely transcend the distinction between therapy and enhancement, as well as not specifying what types of changes can be defined as enhancement, what is desirable and what is forbidden (Savulescu 2019, 321). Since the only criterion here is the well-being of the individual in question, what constitutes enhancement for that individual is a change that improves her well-being.

If we assume that the motivation for space missions must be exceptionally strong—for example, it will be linked to deteriorating living conditions on Earth, or perhaps even a predictable catastrophe—then the solution of increasing our chances of survival in space, as well as increasing our well-being, should be considered an advantage rather than a loss. The decision to accept human enhancement that increases the chances of survival and prosperity is rational, its rejection irrational. In this light, even the possible medical risks associated with its application can be accepted as long as the space mission itself inevitably involves some degree of risk, but above all, it is at least as risky to remain on Earth in light of impending disaster and to consider space missions as a better alternative.

In conclusion, the welfarist concept of human enhancement provides a strong argument for applying human enhancement, at least as long as we agree that our goal is to improve well-being.

Therapy/Enhancement Distinction and Terminological Inaccuracies

There are some fundamental reasons why the distinction between therapy and enhancement must remain imperfect even if it works when analyzing many specific cases. First, it is possible to point to medical practices in use today that are not strictly medical in character, that is, they do not refer to health and are not aimed at restoring the condition of the individual to a normal state. Second, the assumption that procedures interpreted as enhancement are as such morally inappropriate, while procedures with the status of therapy are acceptable, is imperfect. There are medical procedures aimed at safeguarding health that may be morally controversial. And, on the other hand, it is really hard to find anything morally objectionable in enhancement procedures that have clearly positive effects (Cabrera 2015, 50).

Regarding space missions, even if the applied procedures look like forms of even very radical enhancement and thanks to them astronauts obtain functionalities and abilities unknown to any human on Earth, still their consequence is to make these astronauts "normal." Therefore, the consequence of even a radical enhancement will be either therapy or prevention, whose goal is to maintain the astronaut in the earthly norm (that is, the absence of diseases and defects), only with the help of special measures, which on Earth are classified as enhancement, but in space have the status of therapy (or maintenance).

Françoise Baylis points out that even a therapeutic modification is an enhancement, because it always consists in improving the current state, whereas what we understand as therapy and contrast with enhancement should be classified as health-related modifications. Other forms of enhancement are enhancements not related to health. Health-related enhancements, on the other hand, can be divided into treatment and prevention (Baylis 2019).

As Harris points out, what guides human action is the elimination or reduction of harm. Consequently, it does not matter whether we classify a given improvement as therapy or as enhancement, whether it improves a given trait from minimal to average or from average to somewhat or far beyond what is considered typical of the species, if the effect of the change is to reduce harm (Harris 2007, 45–46).

The division between therapy and enhancement also collides with another problem: namely, what we consider to be the norm, and consequently what we may consider to be outside the norm, as well as below it, does not reflect any objective state of nature, but our valuing, constructing, and valuation. This reference to normality and normalcy is particularly unclear in the context of space exploration.

In summary, the prima facie useful distinction between therapy and enhancement faces a number of demarcation difficulties, particularly when it comes to the specifics of understanding human enhancement for space missions.

The Concept of What is Normal in the Space Environment

In discussing the boundaries between therapy and enhancement, the basic concept is the state of what is normal. It is believed that what is supposed to bring us to the point of normality is therapy, and that all those interventions and enhancements that are supposed to take us beyond that point of normality usually have the status of enhancements (Veatch and Guidry-Grimes 2020, 208). But what is the natural state in an extraterrestrial environment? Is a human sent on an interplanetary mission without modification in a normal state? Or is she in an abnormal state? Is a special forces soldier sent on a difficult mission without training equivalent to the level appropriate for such troops in a normal state, even if she still exceeds the average of typical civilians?

Establishing a so-called "normal" state—if it is possible at all—is further complicated by the fact that there are two main philosophical schools of thought that define health and illness differently. According to naturalists, these states are objective and free from valuation, while normativists (social constructivists) believe that these states are social creations and subject to valuation (Veit 2021). Being in space and relating to the specific risk and hazard factors inherent only in space missions complicates this distinction. Naturalism offers a very narrow understanding of medicine as only focused on therapy, hence it is fundamentally opposed to enhancements having non-therapeutic purposes (Rueda et al. 2021). In a sense, normativists may be right, as they receive arguments showing that the same individual (astronaut) on Earth is considered perfectly healthy, while in a different environment (space) he is considered sick, weakened, or below a certain norm (according to earthly criteria).

But also confirmation in favor of their thesis can be found by naturalists, who can point to obvious weakening and exposure of the human organism, for example through high exposure to cosmic radiation. It can be said that from the point of view of bioethical evaluation and meta-ethics, the element connecting the status of health (norm)/disease in space with the evaluation of, for example, autism or disability is that all these states contain elements of both fact and value. As Michael Bury rightly points out, some diseases such as cancer and heart disease can hardly be considered social constructs, while mental health, broadly defined, as well as many other psychosomatic disorders are on a continuum (that is, it is a matter of social relations whether they are seen as diseases and whether they receive treatment) (Bury 2005, 119).

A major challenge for human space missions will be to protect their participants from cosmic radiation. The main difficulties in estimating the risk are related to the limited ability to predict the risk of disease from exposure to cosmic radiation due to the nature of cosmic radiation, and because of the very limited population of those for whom we have knowledge of genetic predisposition to cancers caused by cosmic radiation, the ability to estimate the role played by genetic factors in radiation susceptibility is limited (NCRP Report No. 167 2010). These specific methodological limitations strengthen the argument for the preventive application of human enhancements to protect astronauts from cosmic radiation. This state of epistemological uncertainty related to the limited possibilities of forecasting undoubtedly strengthens the justification for the application of human enhancement, indicating at the same time that unmodified astronauts will necessarily be far below the norm, which in this case is a virtual norm of immunity, but also exposure to cosmic radiation.

In summary, therefore, while human enhancement for space missions is justified by itself in light of the aforementioned definitions of the welfarist theory and Harris's philosophy, the specificity of the space environment, in which astronauts will be far below the norm, provides justification for bioconservatives based on distinguishing enhancement from therapy.

Subjective versus Objective Evaluation of What Is Good and What Is Undesirable

While we will intuitively agree that it is better to have a life without cancer than with cancer, such precise distinctions are more difficult when it comes to the desirability and rationality of proposed applied modifications—whether they are more therapeutic in nature or more akin to enhancement for its own sake (to the extent that such a distinction is possible in a given case). The problem remains even when a given community—for example, a group of physicians—deems a given procedure advisable. This still does not mean that the procedure or its effect is something objectively good or desirable. This is somewhat reminiscent of David Hume's problem of aesthetic value judgment and his concept of expert judgment as the only way out of the subjectivist impasse.

One of the primary conflicts here is between an empirical scientific claim—for example, a demonstrable demonstration that a procedure reduces the risk of contracting a major disease—and the value judgments of a patient who, while appreciating this positive dimension associated with the risk of disease reduction, may suffer side effects that undermine their quality of life (Veatch and Guidry-Grimes 2020, 112). This is a conflict between medical well-being and total well-being.

Another form of this conflict is between ethics based on the principle of duty and the principle of maximizing good consequences where proponents of the first approach may not accept the good results achieved by enhancement applications because they were achieved in violation of at least one of the principles or rules they value.

Consider one of the central principles of duty-based bioethics, the principle of autonomy. The principle of autonomy can both support and limit the moral right to human enhancement. It supports it according to the classical understanding of individual autonomy, where the individual has the right to self-determination. On the other hand, it can limit this right, for example from the point of view of feminist ethics. This limitation could consist in the fact that the individual in principle does not make decisions autonomously, but is determined and constructed by certain social norms and factors. One of these might be the idealized concept of a human being enhanced by radical forms of human enhancement. The argument from autonomy against human enhancement applies particularly to GGE. But human enhancement can enhance autonomy, so even if the decision to adopt human enhancement were dictated by conformist reasons such as those identified by feminist bioethics, it might be worth following them if enhancement were to result in improved autonomy (Schaefer et al. 2014).

Regardless of whether we find consensus on whether a modification is therapy, enhancement or some other type of intervention, an important issue for any modification is its possible coerciveness. Coerciveness overlaps to some extent with lack of autonomy, so the concepts are related. Coercive risk arises when a given enhancement has the status of a positional good and leads to competition. Under competitive conditions, the individual is not formally obligated—for example, through coercive and repressive measures by the state—to apply human enhancement technology. But they may feel strong pressure and fear the loss of their competitiveness, especially if they are a member of a group in which enhancement is widely consumed and can realistically represent an advantage for modified individuals (such groups include, for example, soldiers and students) (Flower 2012).

Thus, there is no doubt that acquiescence to human enhancement can create competitive conditions. Could this risk occur in the context of space missions? One can point to at least two situations in which this might be the case. The first is the situation where in a well-established space base or colony a social hierarchy will develop and human enhancement may serve to obtain privileged or simply better social roles. The second situation, which I discuss when analyzing the rationale for different types of space missions, concerns commercial missions, where the accomplishment of a commercial space mission requires the acceptance of enhancement that an individual, for various reasons, may not want, but which is a condition for obtaining an attractive job.

In sum, the conflict between the subjective and objective vision of what is good overlaps to some extent with the conflict between duty-based and consequence-maximizing approaches.

The Precautionary Principle

As Harris rightly points out, the application of the precautionary principle faces two basic difficulties. The first is related to our ability to predict and our stock of knowledge. Our knowledge is incomplete. Therefore, it is a kind of methodological error to always assume a negative scenario. Second, the precautionary principle in the context of human enhancement requires adopting a certain position on the specifics and nature of biological evolution and the biological status of human beings. As Harris points out, to question the moral legitimacy of human enhancement on the basis of the precautionary principle would suggest that less potential for uncertainty and risk is associated with current human evolution and current and future evolutionary processes that occur without interruption, albeit without intentional interference of the kind associated with intentional human enhancement. Harris emphasizes that the future trajectory of human evolution may be no less dangerous than the dangers and risks associated with the application of human enhancement technology (Harris 2009, 133).

In the context of space exploration, the space environment itself is so dangerous that the possible risks associated with the application of human enhancement do not necessarily outweigh the risks naturally inherent in the very fact of human exploration of space. One should not conclude from this remark that the dangers inherent in a given environment can justify the application of various procedures on humans as part of human enhancement. It does mean, however, that the alternative, that is, sending humans into space without applying any modifications, does not guarantee that it is safer. It can be said to be an ethical situation in which no choice from a moral point of view is appropriate or where all choices are difficult.

While I take the precautionary principle as an obvious principle, I believe that it is difficult to explicitly undermine radical biomedical procedures in the name of the precautionary principle. Context is the key issue here. If the mission is for the survival of humanity, it seems that humanity has no choice but to accept even the most controversial procedures. But if the mission has more trivial goals, perhaps the precautionary principle should prevail.

Human Enhancement versus Human "Nature"

Arthur L. Caplan believes that the essence of the bioconservatives' resistance to human enhancement is their defense of human nature. According to Caplan, the opponents of human enhancement, arguing precisely from the perspective of defense of human nature, are not able to show what human nature actually is. Nor do they show why nature is static despite evolution. Finally, critics fail to show why nature should be a criterion or benchmark for the good or desirable, despite the variability and randomness in biology (Caplan 2004).

As Norman Daniels points out, the concept of human nature is a population concept (Daniels 2009, 31). A change in human nature must imply a change in at least one trait that we regard as one of the characteristics constitutive of human nature, at the level of the population as a whole. Thus, if we wanted to say that human enhancement will change human nature, we would have to realize the following two things.

First, we would have to make a modification of such a trait that can be considered a constitutive or co-constitutive trait for human nature. Thus, it could not be a modification of a trait that could not be considered in terms of a strategic trait from the perspective of human nature. Second, modification of this trait constitutive of human nature would have to be applied on a population level, that is, to all people. As long as the modification of such a trait affects only some people, it is a change in the individual nature of particular people, but not a change in so-called human nature.

In this book, I stand on what might be called an "inert" view. This view holds that even if there is such a thing as human nature, it is only descriptive and is nothing more than a list of current human characteristics or functions. Nothing normative follows from this list (Buchanan 2011b). In conclusion, it is difficult to find justification for the concept of human nature. Thus, arguments against human enhancement that appeal to the concept of human nature cannot be sustained.

Why Biological Evolution Justifies Human Enhancement and Excludes Notions of Nature and Essence

As David L. Hull points out in his foundational article for the critique of the concept of human nature, proponents of the concept of nature or essence in biology are confronted with the fact that the properties they perceive as constituting said natures or essences change over time. Moreover, these characteristics may be shared by almost the entire population at one point in time, while they may almost entirely disappear at another. Such traits are rarely shared to the same degree by all individuals in a population (Hull 1986). Finally, organisms that belong to the same biological species need not share all the same traits, while they must be able to reproduce within the same species.

The argument against enhancement based on the belief that there are immutable natures and essences is untenable. The only potential way to sustain the charge against human enhancement could appeal to the fact that humanity's natural environment is only the Earth. The proponents of such an argument could then invoke that this unchanging common and "natural" element for all humans is the very fact of life on Earth. There would be a specific approach to arguments from an immutable nature, referring to the immutability of where one lives. But the actual immutability of habitability does not preclude the possibility of change, especially when the environmental conditions mentioned by Hull change. And as biological evolution shows, they change constantly and species cross and change different environments.

Hull points out that the only human "nature," the only natural kind in humans, as in all other biological species, is their variability. What constitutes a species is the sharing of a common history by individuals included in a species precisely because of a common evolutionary past, not the sharing of some common essence (Sober 1993; Sterelny and Griffiths 1999).

The critique of the concept of human nature offered by Hull has been questioned. Among other things, critics have charged that Hull's critique applies only to an essentialist notion of human nature, but does not abolish a nomological understanding of human nature. The nomological understanding of nature assumes that humans have properties that naturally evolve in the species *Homo sapiens* (Machery 2008).

According to Edouard Machery's interpretation, we could say that, for example, speech or reasoning are elements of human nature, because they have indeed evolved in humans even if there was once a time when they did not exist, even if they will probably cease to exist in the future, and even though not every particular human being, at least currently, possesses these abilities. Machery adds that part of human nature includes bipedalism, or reacting with fear to unexpected noise. However, none of these traits are either necessary or essential to belong to the human species, nor are these traits necessarily unique to our species alone.

In sum, the perspective of biological evolution indicating continuous change over time both provides a strong justification in favor of human enhancement and questions the meaningfulness of the concept of human nature.

Disenchanting Human Enhancement

Arbitrariness of the Boundary between What Can Be Improved and What Is Forbidden

Once humans have agreed to the constant necessity of improving "nature," it is difficult to point to a rational limit that, in theory, we cannot cross. Any such limit for modifying our "nature" is arbitrary.

Evolutionary Constraints

Opponents emphasize the prohibition of modifying nature. But they themselves admit that our "nature" is far from perfect. This is the well-known problem in evolutionary biology of adaptations that are the product of ancestry and function. Their functionality is often constrained by ancestry, and we would often be able to design a given adaptation better than it actually is as a result of evolutionary processes. Knowing these limitations of nature, why argue so strongly against the demand to improve it, the demand to improve those adaptations that are not too well done? Or perhaps done well, but their shelf life expired in the premodern era, or perhaps even earlier.

Nick Bostrom and Anders Sandberg in their proposed evolution heuristic pose the following meta-question (the so-called evolutionary optimality challenge): "If the proposed intervention would result in an enhancement, why have we not already evolved to be that way?" (Bostrom and Sandberg 2009, 378). Bostrom and Sandberg give several reasons why this might be the case.

First, the evolutionary process is a process of trade-offs between different interests and demands. Second, humanity's belief about what is attractive, what should be improved or even just created in the human organism, often falls short of the specifics of how evolution works. Third, the idea of human enhancement as it is understood today provides humanity with unique tools for rapid change that are not available to evolutionary processes. As Bostrom and Sandberg point out, that does not mean we are better than evolutionary processes. What it does say, however, is that there are situations in which evolutionary processes are unable to generate a given trait or functionality, while a single man-made tool—in this case, biomedical technology—is able to make the change (Bostrom and Sandberg 2009, 378–380). This is related to cellular- and DNA-level constraints and corresponds to the concept of cellular liberty postulated by Mason (2021). This idea has obvious implications for the context of space exploration considered here. If humans are not even sufficiently adapted to the modern environment on Earth, then they are even less adapted to the completely different space environment. Since we cannot possess certain traits and adaptations despite living in an earthly environment, much less can we expect to evolve adaptations to a space environment in which we, as a species, have never lived. Knowing that evolution is not a perfect process and that many elements could be improved, we can on this basis formulate an obligation to improve what evolution, due to various constraints, could not make perfect.

An Argument from the Perspective of the Genetic Lottery

The genetic lottery argument is a strong argument for applying human enhancement. People are born with different genetic endowments. These differences mean that some may have certain predispositions that others do not, as well as being able to perform certain activities better than others. This is an example of an unfair distribution of abilities and potentials caused by the genetic lottery that can be corrected by human enhancement (Savulescu et al. 2004).

It is worth keeping in mind John Rawls' explanation of natural distribution. As Rawls points out, the mere distribution of traits or abilities cannot be judged in moral terms of justice or injustice. But what can and should be evaluated in moral terms is our response to that distribution, or more precisely, the response of institutions when the effects of a natural distribution deemed unjust can be nullified or at least reduced (Rawls 1999, 87).

The argument over the naturalness of the process, declaring the genetic lottery morally acceptable or at least morally neutral, does not explain why inequalities between people caused by their genetic differences—assuming that such differences can affect career and wealth development as well as the social position of individuals—are not a moral problem, while possible inequalities caused by human intervention are a moral problem for opponents of human enhancement. Thus, while one could agree that human enhancement in some sense may limit the open future of the modified child, it is worth remembering that the absence of human enhancement does not make the future of the unmodified child open. There remains the genetic lottery, which, at least insofar as we are at least partially determined by our genes, can also limit the degree of openness of our future to no less a degree than human enhancement (Sandel 2004). The bioconservatives seem to have run into a dead end here. If they defend the genetic lottery against human enhancement on its own merits, then they should equally oppose medicine as our response to nature's adversities and failures. If, on the other hand, they are defending the gene pool from greater corruption threatened by human enhancement, then they should similarly renounce medicine all the more and, admitting openly to social Darwinism, allow only the strongest that survive without medical assistance to remain in the pool.[1]

In summary, the issues discussed in this section such as the lack of clarity between what can be enhanced and what should be prohibited; the imperfection of biology and evolution, and the genetic lottery provide strong arguments for human enhancement.

Arguments against Human Enhancement

Much of what has been said above regarding critics of human enhancement boils down to the conclusion that human enhancement is not fair (The President's Council on Bioethics 2003; Callahan 2003). Human enhancement has great potential to exacerbate inequalities, but these exist regardless of the availability of biomedical modification measures. However, human enhancement also has great potential to redress inequalities (Alonso et al. 2020). If certain traits are, at least to some extent, heritable, only biomedical interventions are able to eliminate the trait considered to be a lack or defect, at least to the extent that it is determined by biology and not by environment.

The authors of the report of the President's Council on Bioethics rightly emphasize that the approach and ethical evaluation of various forms of human enhancement will depend to a large extent both on the availability of ever new means of carrying out such modifications and on public opinions, trends, and sentiments. As examples, the authors point out that the decision to apply performance-enhancing drugs may depend on the degree of competition in a given culture, whereas the decision to apply modifications to one's own children will depend on whether such modifications are applied by other parents (The President's Council on Bioethics 2003, 275–276). This, then, is not so much a direct unfair indictment of human enhancement technologies as an indirect indication of the risk of producing a situation that will produce broad unfair consequences. In contrast, it is worth noting that this objection, however real, does not seem to apply to the practice of applying human enhancement to space missions. This is due to the specific justification of human enhancement in space, which is related to health issues, even if it is applied to healthy individuals and even if in practice it will consist in providing them with new abilities or features not possessed by the rest of society.

Another, much more important reason why this objection does not apply to space missions is their exclusive nature. If these enhancements are applied only to astronauts and remain rationed, there is no risk here of their market release and creating an unfair effect. Preventive and therapeutic enhancements applied solely to space missions have little chance of creating a public desire for them and thus influencing public desires and tendencies.

Human Enhancement as a Threat to Human Dignity

Rebecca Roache and Julian Savulescu propose a thought experiment to understand what constitutes the core of bioconservatives' resistance to human enhancement. To this end, they propose that a modification be made to enhance or generate those qualities that are traditionally valued by bioconservatives, such as wisdom, human dignity, and respect for the given (Roache and Savulescu 2016, 150–151). Bioconservatives point to the danger of dehumanization, of treating human beings as things, of risking the deprivation or diminution of human dignity.

We can have many different concepts of human dignity. But if we were to adopt here the Kantian understanding of dignity, according to which dignity is moral conduct, which in turn is conditioned by the possession of freedom, that is, the ability to act on the basis of reason, then human enhancement by biomedical means can only improve our ability to be guided by reason (Chance 2021). Thus, Roache and Savulescu's argument can be read as a perverse argument in favor of biomedical enhancement directed at its conservative opponents, showing that the effect of such modification will be only that which is morally good (i.e., the values cherished by conservatives), and what is new is the means of achieving it.

Social Objections to Human Enhancement

Finally, it is worth considering social objections to human enhancement in relation to space exploration. The social objections can basically be boiled down to the following idea: human enhancement technologies may become a market good, which will be available for the few (the wealthiest), and thus lead to inequality and discrimination.

Human enhancement applied for space exploration can become a marketable good and raise the same social issues as modifications applied on Earth, provided that the population living in space generates a social hierarchy that makes human enhancement marketable. But for this to happen, highly advanced technology far beyond the needs of sustaining life in space is not enough. An appropriate legal and institutional environment is also needed. And here I see a great unknown, which at the same time is a great opportunity for humanity to avoid the mistakes of the past, if we believe in moral progress. But if so, then a space colony is best suited to play the role of embodying moral progress in its purest form, that is, building a just society from the ground up.

Is the Invasiveness of Biomodification an Argument against Its Use?

One of the reasons why the concept of human enhancement is considered morally controversial is the invasive nature of such "radical" forms of modification as gene editing and implant placement. The very concept of invasiveness, however, is troublesome to maintain. Widely used and accepted vaccinations are themselves invasive. But it is not the degree of invasiveness that characterizes surgery. It is worth noting here the surgical procedures, which are, after all, a highly invasive activity. Their most radical form is transplantation and the application of artificial organs. Despite this, they are widely accepted and recommended in order to save lives. Since, therefore, we come to the point where we find that practices of various kinds with even a high degree of invasiveness are quite common and basically no one questions them—moreover, they seem to be a major social good desired by society as such—then we have a rationale for concluding that it is not invasiveness as such that is the problem, but something else.

The mentioned examples of invasive procedures refer to life-saving medical procedures. But—if we have assumed that invasiveness as such is not an issue—does the purpose of the procedure considered invasive become an important parameter here? Perhaps we consider said invasive procedures ethically uncontroversial only because they are always accompanied by a medical purpose. Such coincidence—not only in this but also in other cases—can distort the ethical evaluation because it makes it difficult to isolate a factor in itself if it is always associated with another that is acceptable, as in the cases discussed here, where a potentially controversial procedure that is invasive in itself is always (in the medical cases discussed here) associated with a universally acceptable rule, namely, saving life.

But treatments that we would rather not see as therapeutic in themselves, such as aesthetic medicine procedures, are also not considered controversial despite the fact that, such as in breast implants, they are no less invasive than life-saving surgery. The therapeutic component here—if only for mental health—can be substantial. We have a situation where there are invasive treatments that are qualitatively no different from medical treatments in terms of invasiveness, but which are not at least prima facie therapeutic and yet are not controversial. Nevertheless, invasiveness is a factor that has an ethical weight. There is some rationale for the claim that it may be better to avoid invasive procedures and opt for noninvasive or less invasive procedures when available.

In conclusion, some of the most popular arguments against the application of human enhancement such as the argument that human enhancement is unfair; human enhancement as a threat to human dignity; social objections, and the issue of the invasiveness of human enhancement, lack sufficient justification.

The Real Case against Human Enhancement

I reject the allegations against human enhancement presented above. I consider the following to be the only legitimate objections. As Buchanan rightly points out, implemented enhancement as he understands it can go wrong. This is the case when, in spite of the applied modification, the assumed goal is not achieved, as well as when the introduced change leads to a worsening of the situation (Buchanan 2011, 24). It is worth considering the possibility of both situations occurring in the space environment.

When the Goal Is Not Achieved

What I mean by the goal of enhancement is a change in function. Such a change might be an increase in resistance to cosmic radiation or a reduction in the negative effects of prolonged exposure to altered gravity. Let us imagine that the planned change in functions described above has not been achieved. Since these changes must still be carried out on Earth, in the prelaunch phase, it is difficult here to point to any moral controversy of human enhancement, which did not bring the expected results. We assume that neither the work on human enhancement, which ultimately proved ineffective, nor its application itself caused harm to anyone, taking into account both the loss of public funds and the possible discomfort of those undergoing the modification.

A bigger problem arises when some type of enhancement is planned to be applied during the realization of a mission, at some stage of its realization, or just before the realization of some special task. We can theoretically imagine that a given type of biomodification can be applied only in the conditions of a space mission and only shortly before its specific phase, which requires its implementation. Let's assume, then, that for technological or medical reasons or both, carrying out such enhancement on Earth is not possible. If, however, it turns out that its application in space did not produce the expected results, the moral evaluation depends on whether the persons carrying out the mission are able to find out about the failure of the human enhancement application immediately or only on the basis of the health consequences for the person subjected to the ineffective modification. The moral evaluation of the second situation is more severe. However, in both cases it cannot be said that the biomedical procedure itself turned out to be dangerous (because, de facto, what brought danger or forced to cancel the mission was the lack of human enhancement, not its application).

When the Modification Makes Us Worse Off

This option is more amenable to formulating a clear moral judgment than the scenario discussed above. The harmfulness of human enhancement effects will be greater if they become apparent only during the course of the mission. It may be that various types of on-target and off-target effects will occur, which will not only adversely affect health but also may lead to the death of the modified individual. They may also lead to the catastrophe of the entire space mission. However, this is definitely a clear-cut situation in terms of moral judgment. The alternative here is a decision to postpone the mission or to replace the human mission by an automated mission.

Despite the fact that we have good reasons for equating enhancement with improvement, there are also good reasons not to do so, or at least not for every kind of enhancement. Chadwick and Schüklenk discuss in this context—enhancement as improvement—the issue of its desirability. They point out that not every enhancement (improvement) is desirable. What is important is not only the goal of a given enhancement, but also the success in achieving this goal with the help of a given modification (Chadwick and Schüklenk 2021, 162).

In conclusion, the only basically justified arguments against the application of human enhancement refer to the impossibility of achieving the planned goal, as well as to the creation of negative effects of the applied modifications. These are quite unambiguous moral situations, which in essence can be reduced to the problem of guaranteeing certainty and predictability.

Disability in Space Missions

There is no doubt that for many conditions considered disabilities, modification of current environmental conditions would make these conditions either neutral features or even warrant reasonable accommodation (Barclay 2016, 84). This is the type of argument, to some extent valid, that often opens a space for discussion of social injustice and inequality and exclusion. Let us look at how this problem of exclusion is solved in space. I distinguish at least two contexts here. One is the existence of environmental barriers, the other is the existence of social barriers supposedly justified by environmental ones and their impact on the selection of candidates for space missions.

The comparison to military service may be useful here. Just as the selection of candidates for the army places a premium on those in good health and physical condition, the selection of astronauts requires similar, or even greater, qualifications. If we look at the specifics of space travel, the required activities including extravehicular activity or the range of tasks awaiting future astronauts after landing on Mars, it can be surmised that certain types of disabilities may either prevent or exclude the involvement of people with disabilities. However, such candidates should not be excluded, and mission organizers should ensure that they are allowed to participate in missions after determining that a given type of disability, with the technology currently available, will not prevent mission tasks from being accomplished.

Two issues emerge here, one supportive and the other detrimental to people with disabilities. The former is that there may be forms of disability that give people with disabilities an advantage in the space environment, or are simply useful. Examples include blindness, given that cosmic radiation is particularly harmful to vision. Another example is lower gravity, which can abolish the difference between able-bodied and disabled people with respect to locomotor organs. There are probably other types of disabilities that either give disabled people an advantage over nondisabled people in space, or make a given disability unproblematic from the point of view of the effectiveness of a given task.

The second controversial issue, on the other hand, is the question of whether there is a moral right to exclude disabled persons from participation in space missions or to forcefully subject them to modification. In the first case, the morality of such a prohibition might be justified by the mission organizers' proving that a person with a given type of disability is unable to perform a given task in space, which is not only strategic (i.e., cannot be abandoned), but also cannot be performed by another mission participant (because, for example, it is the kind of task that every astronaut must perform). In this situation, environmental considerations provide justification for prohibiting the participation of people with disabilities.

The alternative is to require a modification, in this case undoubtedly of a therapeutic nature, that nullifies a given characteristic that is considered disabled and not essential to the space mission scenario. If a candidate with a disability accepts such a criterion, the moral problem disappears. But if the candidate finds it controversial, we are back to the moral situation described above, where mission organizers must prove that the modification is necessary.

Recognizing the primacy of subjective well-being and subjective self-assessment can lead to a somewhat paradoxical situation in which the environmental context and the attempt to establish any objective criteria loses its justification (Barclay 2016, 84). In the context of space missions, the only test to verify the veracity of this claim is the aforementioned case-by-case analysis (of each individual candidate and specific type of disability) and whether a particular disability may prevent the necessary and strategic tasks that must be performed by that particular candidate. Otherwise, the exclusion of people with disabilities will be another example of their arbitrary exclusion and injustice known from Earth.

Finally, I would like to address an idea proposed by some philosophers of space exploration who oppose the idea of mandatory human enhancement for space missions in the name of justice and equal access (Schwartz 2020). It is worth distinguishing two issues here. One is the rightly emphasized value of justice and equal access, but the other is the real possibilities and specifics of a particular situation. Let us imagine a situation in which participation in a given space mission, regardless of its purpose, requires the application of radical human enhancement for purely medical, therapeutic reasons. Otherwise, people taking part in the mission are exposed to serious health problems or even death, not to mention the risk for others through the inability to perform strategic tasks during the mission.

In conclusion, the concept of human enhancement for space missions does not pose a threat or problem for people with disabilities. Possible restrictions could be allowed only after proving that there are certain technical or environmental constraints that either preclude the participation of people with disabilities in space missions or oblige them to undergo human enhancements.

Human Enhancement as Our Moral Obligation

I believe that the future of humanity without human enhancement will be bad rather than good, and conversely, the application of human enhancement will bring more good than bad. Consider some fictional scenarios of a future world in which human enhancement is prohibited.

The first scenario. Human enhancement can be used for disease prevention. Imagine being able to make such changes to the genome of an embryo that it will be immune to a certain type of disease—or perhaps several types—which, when cumulative across a population, has adverse effects. These effects could include the suffering and death of sick people, the correlated social burden, and the economic expense. However, we live in a society in which we are not allowed to inflict human enhancement, especially one as radical as GGE. The cost paid for being faithful to our moral principle, which is the idea of the immutability of our nature—as absurd as it sounds from the point of view of genetics and biological evolution—is the suffering and death of the sick, as well as the aforementioned attendant costs and losses.

The second scenario. Let us imagine that a way to mitigate the negative effects of climate change is biomedical moral enhancement. Imagine that there are tools to biomedically modify our morality in such a way that our decisions and actions do not harm the environment. The goal of such biomodification would be to make the concept of climate change, the negative effects of global warming, and overpopulation a close idea to us, as close as one's desire to do good for one's genetically related relatives. Imagine if biomedical moral enhancement led to mechanisms that worked the same way as kin selection. But our deep-seated moral intuitions make the concept of moral bioenhancement incompatible with our principle of the immutability of our nature. We therefore admit that it is not in our nature to act pro-environmentally, nor to act for the good of the population as a whole, which overlaps quite a bit with pro-environmental action. Since we are unable as a population to change our habits regarding energy consumption as well as reproduction, we are inevitably headed for disaster.

The third scenario. Let us imagine a near future where humanity is forced to leave Earth and settle somewhere in space. Let us assume, then, that the future conditions of life on Earth are possible, where life in space will have a better quality of life, or will be more worth living, than life on Earth. Consequently, living in a space colony will offer the chance for a greater level of happiness than remaining on Earth. But let us assume that the prerequisite and necessary condition for participation in a space mission is human enhancement. Since we live in a society of moral bioconservatives, this possibility is not accepted by us.

In the scenario described here, humanity may be aware of the risks of continuing to remain on Earth and is aware of the need to evacuate, but its energy and efforts are focused on even more intensive development of space technology and medicine. These efforts do not even mention such an option as human enhancement. As a result, humanity is annihilated, however not completely, and the survivors have lives barely worth living. Alternatively, instead of a single catastrophe of global proportions, humans experience steadily worsening living conditions, and perhaps know that life in a space colony they did not have time to build would be a life more worth living.

And now, for a change, imagine a world in the not-too-distant future where human enhancement is widely accepted and used just like education or strictly therapeutic methods. In such a society, the appearance of new diseases is not a problem as long as they can be prevented by somatic or germline gene editing. Humanity simply embraces the available means, seeing nothing immoral in them and instead seeing the positive effects of preventing or even eliminating disease.

At some point, humanity will realize that the only way to confront climate change, and incidentally perhaps solve the overpopulation problem, is to apply moral bioenhancement. Let us assume that it would be medically and technically feasible. Humanity on a global level chooses to undergo such modification so that pro-environmental and also pro-population behavior becomes intuitive. Consequently, no one even recognizes that environmental concern involves giving up every day conveniences and requires perhaps a major modification of habits.

Finally, space colonization will not be hampered by the adaptive limitations of the human body. Under the quite realistic assumption that the only obstacle will remain the ability to effectively protect human health and life from the harmful effects of cosmic radiation, and that the only alternative for a long time will be human enhancement alone, humanity will be ready to pursue long-term space exploration leading to the establishment of space settlement.

I believe that the human enhancement ban interferes with and calls into question the roles and functions we assign to science in general and medicine in particular. Are their roles not to serve human beings, to enhance their quality of life, to remove problems and risks? If so, then human enhancement fits into this set of functions ascribed to science.

In sum, although I refrain from concluding that human enhancement can be understood in terms of moral obligation, I believe that those possible future scenarios that assume widespread use of human enhancements better serve well-being than alternative scenarios that exclude the use of human enhancement.

4

Germline Gene Editing and Embryo Selection for Future Long-Term Space Missions

Introduction

The subject of this chapter is genetic modifications focused on adapting humans to the space environment. Today's context for discussing somatic and germline gene editing (SGE and GGE, respectively) is set by three social and cultural phenomena. First is the He Jiankui experiment. Second is the still persisting moratorium on GGE, which in principle forbids this procedure and considers its application only in special medical cases. Third, we can observe strong criticism not only of GGE but also of embryo selection, which is often referred to as eugenics. The thrust of this chapter is as follows. If we are criticizing human enhancement in general, and through GGE in particular, it is worth considering what we are really criticizing, what element we are concerned with, perhaps the breaking of a rule, perhaps the effects, perhaps something else. This is important because in rejecting the arguments of bioconservatives, usually what is criticized in human enhancement is misidentified, does not exist, or exists in many other nonbiomedical enhancements that humanity has been applying for hundreds or thousands of years. It is worth keeping this rule in mind with regard to criticism of GGE, because it may be that what we accuse GGE of doing is being done in other areas perhaps to an even greater extent. The hope of obtaining very good consequences creates, in my opinion, a strong pressure on opponents of GGE to provide convincing arguments for not applying such procedures.

An important type of procedure is embryo selection, which, although it is not an intervention in the structure of the embryo in the same way as GGE, can nevertheless serve in a certain sense the purposes of enhancement and not only of therapy or prevention, since, in the end, the selection of a healthy embryo leads to the same consequences as the enhancement of one that is below what is currently considered, in a given context, to be the norm. Therefore, an important part of this chapter will be a reflection on embryo selection as an important form of so-called designer babies, which is also more realistic from a scientific point of view—because it is already being carried out, for example, within the framework of preimplantation genetic diagnosis and testing (Klitzman 2020).

The Bioethics of Space Exploration. Konrad Szocik, Oxford University Press. © Oxford University Press 2023.
DOI: 10.1093/oso/9780197628478.003.0004

SGE

SGE is an editing of nonreproductive cells, different than gametes (egg and sperm), zygotes and embryos which are the sex (germ) cells (NHGRI, n.d.). I consider SGE to be essentially an uncontroversial procedure primarily because SGE makes nonheritable changes to DNA (Baxter 2021), despite the fact that I do not regard the fact of heredity as a controversial feature. SGE is a promising form of gene therapy that can cure even more than 10,000 human monogenic diseases (Memi et al. 2018) and is devoid of both the technical and ethical controversies inherent in GGE (Chang 1995). SGE fits well into the specifics of modern medicine, especially in the context of treating such serious diseases as cystic fibrosis and sickle cell anemia (Stock 2002, 39). SGEs affect only the organism on which they are applied.

One gene-editing technology is CRISPR-Cas9 (Klug et al. 2020, chapter 17). It is a technology that has advantages over other forms of gene editing mainly due to its relative ease, low cost, as well as precision (Travis 2015). CRISPR-Cas9, unlike other forms of gene therapy, does not involve inserting a new gene alongside a defective gene, but relies on genome editing by deleting, correcting, or replacing the mutated gene, and is more efficient and easier to design than, for example, zinc-finger nucleases (ZFNs) and transcription activator-like effector nucleases (TALENs) (Klug et al. 2020, 475, 476). CRISPR-Cas9 can be applied to both SGE and GGE. SGE by CRISPR-Cas9 has, and probably will have in the future, many practical nontherapeutic applications which can be attractive to many age groups (young and old) as well as professional groups (athletes or soldiers or employees of various services where fitness is important, and, of course, astronauts). CRISPR gene editing has many therapeutic applications in both humans and animals (Luo 2019).

Let's stay with CRISPR-Cas9-mediated SGE for now, which are either preventive or therapeutic. The main application areas include, but are not limited to, the treatment of various types of genetic diabetes, sickle cell disease, diseases that cause blindness, Duchenne muscular dystrophy (Congressional Research Service 2018), HIV (see: first patient cured of HIV, Timothy Brown, with ZFNs (Klug et al. 2020, 475)), and use in clinical trials for the treatment of cancer and genetic diseases of the muscle, blood (leukemia, hemophilia), liver, but also hearing loss, Alzheimer's, Parkinson's, amyotrophic lateral sclerosis, cardiovascular disease, and muscular dystrophy (Klug et al. 2020, 343, 468). Of the more than 2,300 clinical trials with gene therapy worldwide, just over half are cancer treatment trials (Klug et al. 2020, 343, 468–469).

SGE can be performed in a precise manner both ex vivo and in vivo (Tozzo et al. 2020). For example, one treatment for HIV is to remove T cells from HIV-positive humans, then ex vivo use ZFNs to disrupt the CCR5 gene, and finally introduce the modified cells to patients (Klug et al. 2020, 475). CRISPR-Cas9 has been successfully used to edit human, monkey, mouse, rat, and zebrafish cells, among others, showing greater efficiency for dividing cells than nondividing cells (Meneely 2020, 285).

Genome editing involves adding copies of a functional gene to somatic cells, allowing nonfunctional alleles to be replaced with functional alleles (Meneely 2020, 289).

Currently, only SGE is allowed in gene therapy, so-called somatic gene therapy, where the effects of the intervention affect only one patient and the criterion of informed consent is met for either the patient or the family (Klug et al. 2020, 343, 480). The criterion of affecting only one individual is undoubtedly important from the point of view of bioethics, because, regardless of the effects, even positive ones (ignoring possible negative side effects like off-target effects), we avoid the situation when we affect other individuals despite their consent/no consent and lack of knowledge.

In conclusion, SGE is a promising, effective, and ethically uncontroversial procedure that has been successfully used for therapeutic purposes.

Arguments against SGE

In principle, all of the objections to human enhancement discussed in the previous chapter can be applied to SGE. However, focusing specifically on SGE, it should be noted that the main argument against SGE relates to risk and uncertainty. As long as we are not sure how the body will react and what side effects the SGE application will bring, we should abstain from such an experimental therapy. The problem is still the risk of off-target effects. In the case of methods earlier than CRISPR, such as gene therapy with retroviral vectors, there have been situations when the introduction of a retrovirus activated neighboring genes, leading, for example, to leukemia in children who have been successfully cured of immunodeficiency disease (Dale and Schantz 2007, 350–351).

It is also worth bearing in mind the increasing knowledge of our genome and perhaps the growing popularity in the future of SGE applications for non-therapeutic purposes, as well as the danger caused by the knowledge of the genetic sequence. The mere proliferation of genetic tests and genomic sequences can lead to a kind of genetic discrimination that is particularly likely to manifest itself in areas of life such as employment and medical insurance (Klug et al. 2020, 343, 520).

Such genetic discrimination may be particularly justifiable in the context of space missions, albeit with the proviso that the party covering the costs will be the mission organizer (except perhaps for tourist flights). I see no moral impediment why an astronaut would not agree to an SGE application that would increase their resistance to cosmic radiation, reduce the rate of bone density loss, and slow down muscle atrophy. A candidate to be an astronaut has no basis for considering such requirements to be discriminatory so long as she voluntarily applies to work in space and humanity has not produced alternative countermeasures. But I also do not think that the presence of alternatives changes the moral situation. If SGE turns out to be faster to apply and cheaper than nonbiomedical countermeasures, that is a compelling argument for treating SGE as a mandatory procedure, or at least recommended if time and cost optimization is an important factor in a given context and for a given space agency.

The moral status of forcing the application of SGE for commercial space missions and colonization (space refuge) missions may be slightly different. With respect to the former, I believe that no clear criterion can be established as long as we do not determine what the attractiveness of this work is compared to other offers, what the risks are, as well as whether the modification serves more therapeutic purposes or the purposes of so-called enhancement, which in this case would mean serving to increase the productivity of the worker. One option is to consider this a matter of voluntary workplace choice, another option is for the employer to fund the SGE and collect the cost of the SGE from the employee's future income. Finally, another option is where we recognize that our approval or disapproval of requiring SGE for commercial purposes depends on the usefulness of commercial space exploration to humanity as a whole.

The biggest controversy, however, will be the requirement of nontherapeutic SGEs for colonization missions treated as saving humanity, not just a simple derivative of commercial space exploitation. The requirement to apply SGE for therapeutic purposes will not be controversial as long as we do not consider it controversial to wear seat belts or a helmet where seat belts or a helmet are required for safety reasons. However, as this is a special requirement, which does not exist in the case of flights by passenger aircraft on Earth, the obligation to finance it should be ceded to some institutions, and not to individuals who are to take part in the colonization mission.

In the case of an SGE obligation of a nontherapeutic nature, but one that will nonetheless be considered highly relevant to the operation of the mission, in keeping with the current space agency policy of assuming that each astronaut must remain productive for the entire duration of the mission, it is necessary to dividend what the modification is needed for, and whether exactly the individual should receive it. It is possible, since we are talking about SGE for enhancement and not therapy, that different individuals will receive different SGE. For what about a situation in which every participant in a colonization mission must be compulsorily subjected to such a modification for their future function in the space colony? Such a choice will determine their role and social position, and it is not clear whether transfer between social roles will be possible. Perhaps it will require undergoing another SGE that nullifies the previous SGE obtained on Earth. One can see here that such a concept, even if it were indeed justified by a morally good concern for the well-being of the human population in space, gives rise to an enormous, perhaps entirely unplanned, but perhaps also impossible to eliminate, risk of class divisions, hierarchies, and potential social tensions and struggles.

In conclusion, the main objections to the SGE are related to safety issues, primarily the risk of off-target effects. Some reservations—although highly dependent on the context and circumstances considered—may also be raised with regard to the social effect of SGE application, especially in the context of the labor market.

Arguments for SGE

Genome editing can be seen as a more precise type of gene therapy that allows for the treatment of monogenetic diseases by fixing the genetic mutation that causes the disease and replacing it with a sequence that does not cause the disease. It is therefore a way to restore DNA to its normal, nonmutated state (United States Senate 2017, 6).

In addition to this restorative function, there are also a number of possible applications which we tend to associate intuitively with enhancement rather than therapy. SGE can be used to improve muscular abilities and increase fitness and performance (President's Council on Bioethics 2002).

SGE may be the main type of human enhancement applied to space missions. Its potential applications could be in adapting the individual to all those threats in space that are currently the most disruptive (Mason 2021). But one can also consider potential applications concerning the enhancement of behavioral and cognitive adaptations, which, although not having therapeutic or preventive functions in themselves, seem to be essential for survival in such a harsh and isolated environment as the space environment. It is worth recalling the fact that one of the three major medical challenges in future long-term space missions is behavioral adaptations. The identification of genes that enhance certain behaviors and emotions, considered desirable for the safety and success of a space mission, may lead to the obligatory application of SGE for moral and cognitive bioenhancement purposes.

In sum, SGE is considered as a major type of human enhancement for future long-term space missions, with great potential for performance enhancement applications.

GGE

GGE includes editing of gametes, zygotes, and embryos. One of the methods used in GGE, for both viable and nonviable human embryos, is the aforementioned CRISPR-Cas9 (Ma et al. 2017). Ongoing studies have targeted diseases such as beta-thalassemia, hypertrophic cardiomyopathy, and Marfan syndrome, the latter applied to viable embryos with a repair rate of 89 percent and no off-target effects (Congressional Research Service 2018).

GGE is considered a simpler procedure than SGE. This is mainly due to the fact that with GGE we can easily modify the targeted genes from the beginning as a whole, since the targeted genes are located in each cell, whereas with SGE we have to target a particular gene in a particular place to make it active in the right place and in the right way (Stock 2002, 39). The advantage of GGE over SGE is the greater efficiency of GGE (Meneely 2020, 298). GGE can involve both sperm and ova as well as the early embryo. In all cases the genetic changes will pass on to future generations. These generations, however, will live under different environmental conditions in which the changes applied in one generation will not necessarily be compatible with those of subsequent generations born with the same modified trait (Meneely 2020, 289).

The first genome editing conducted on nonviable embryos in China in 2015 and 2016 resulted in failure. Only four out of seventy-one embryos achieved the intended HBB gene modification, while off-target effects, that is, modification of genes other than the intended ones, were reported in many cases. The same was true for the 2016 study, when of twenty-six embryos undergoing CCR5 gene editing, only four completed the intended editing. In contrast, a 2017 study on viable embryos, also in China, was quite successful in treating disease-causing mutations. Nonetheless, the problem with applying CRISPR-Cas to GGE remains both the risk of unwanted changes to the target gene and the risk of off-targets (Klug et al. 2020, 478). The medical risks are high, especially for long-term effects that may manifest in the modified individual and even in their offspring (Turocy et al. 2021).

In summary, GGE refers to a modification of sex cells that is inherited to the next generation. Currently, this procedure is still full of medical risks.

Arguments against GGE

Ethical Reservations Related to GGE's Failure

One risk is when genome editing results in mosaic genes, when only some of the cells for the same gene are edited. This can result in a situation where only some of the cells in the embryo are modified. Another risk is off-target effects, which are situations where genes other than the intended ones are edited. In humans, this risk is compounded by polymorphism, which makes it difficult to predict off-target effects, which are distributed unevenly in different people (Meneely 2020, 290). Pleiotropy only increases the degree of unpredictability (Meneely 2020, 291).

These are strong arguments against the application of GGE, and I treat them as belonging to a single set of risk and safety arguments. I consider meeting the safety criterion of the procedure as an unavoidable, necessary condition. One may wonder, however, whether there are scenarios in which we would be ready to apply GGE despite the uncertainty of the dangers it may bring. When it comes to the context of space missions, we can imagine a scenario in which we recognize that we need to reproduce because we need replacement generations. Suppose we have calculated that in a given year the population in space needs to conceive a certain minimum number of new children, and that everyone needs to be given a GGE for environmental reasons (without a GGE the child will be born in worse shape than with a GGE). We accept the risk of complications after GGE, but otherwise know that it will be less than without GGE. This is a context of calculation that is different in space than on Earth, because on Earth we could abort such an embryo, whereas in space we cannot, because it will prevent the replacement of generations and the success of the colony at some point in its future existence.

Ethical Reservations Related to the Success of GGE

Success in GGE means that there are no off-target effects, that is, no unplanned genes have been edited, and also means that all cells in the embryo have been edited. The problem is the issue of consent from future children and the risk of possible resentment years later from the parents.

I believe that many nonbiomedical changes parents make to their children have lifelong consequences and are likely to be irreversible. It is also unclear whether their balance actually benefits the child. We can already point to the existence of many different social, economic, and cultural norms and criteria that cause a rat race. There is no reason why GGE will be qualitatively different, however it may make implemented changes in one generation no longer in this race to catch up for that generation if the next generation gets an improved version of that modification (compare Sparrow's thought experiment about the so-called yesterday's child (Sparrow 2019)).

Another problem is the difficulty in predicting the phenotypic effects of applied genetic changes (Meneely 2020, 290–291). But I assume that here we should simply wait for advances in genetic knowledge that will verify which phenotypic traits are inherited, and to what extent.

An Argument from Development Potentiality

This is one of the most popular attempts, along with the belief in the moral status of the embryo, to defend its right to exist and its inviolability. This argument is not so much about the status of the embryo as such, but about its future after birth.

And herein lies the fundamental difficulty about the potentiality of rights. Opponents of this argument like Devolder point out that the egg is neither a chicken nor an omelette, and as long as it is an egg, it can become either. Or, already somewhat more ironically, reducing the argument from future and potentiality to absurdity, she points out that from the fact that each of us will die, it does not follow that we should already be treated as dead (Devolder 2019, 253).

Another example of reducing the aforementioned argument to absurdity is its scope. Devolder rightly points out that, under certain conditions, any cell of the human body that has the potential to develop into a human being will thereby have to be accorded the same moral status as the embryo currently has according to the argument from potentiality of development. This special case is cloning. Since every cell can be cloned into a full human organism, every cell therefore has a moral status equal to a human being (Devolder 2019, 255).

GGE seeks to correct what natural processes, and to some extent evolution itself, have failed to do.

Open Future and Autonomy

While GGE tends to be associated with closing off an "open" future and reducing autonomy, it is often the lack of GGE and human enhancement that can close off an open future, and at least reduce the number of possibilities (DeGrazia 2012). This problem is pointed out by Lewens, who says that enhancing certain abilities and characteristics in the future child will increase their chances of success rather than reduce (Lewens 2015, 202–203).

Does GGE in the context of space missions make the future more or less open? Imagine humanity pursuing advanced spaceflight. Suppose that in order to pursue a certain type of space mission, a GGE applied by parents to a future child on Earth is required. Without it, that child will never be able to participate in a given type of space mission. What decision should the parents make? It depends on whether a particular GGE will affect the quality of life and the range of opportunities on Earth. If so, then there is a strong rationale for not applying a GGE to a future child since, at least under certain conditions, such a GGE may force the child to emigrate from Earth (because the available alternatives will be less attractive than a space mission).

But if the GGE does not affect the range of future possibilities on Earth, then there is a strong rationale for applying the GGE. Because participation in a space mission that is considered attractive is clearly an added bonus that a future child will be deprived of if the parents choose not to apply the GGE. However, this depends on various conditions, and this thought experiment can be complicated in many different ways. Referring back to Sparrow's experiment on "yesterday's child," it can be said that, in a sense, a child born without GGE, which is essential for space missions, is a loser child compared to those children of the same and other generations whose parents have applied GGE to them.

Arguments for GGE

Procreative Beneficence

The principle of procreative beneficence, formulated by Julian Savulescu, states that "couples (or single reproducers) should select the child, of the possible children they could have, who is expected to have the best life, or at least as good a life as the others, based on the relevant, available information" (Savulescu 2001b). Other formulations of the same principle are "the maximizing constraint" and the "best interests of the child" principle that allow a particular intervention to be judged bad when it results in a child being born worse off than if it had been born without the intervention (Savulescu 2001a).

Savulescu, along with Guy Kahane, adds that the procreative beneficence principle does not explicitly say exactly what choices should be made, which is particularly relevant to the disability case they are discussing. What is the guiding principle here is the principle of the welfare and well-being of the future child and what will give them an advantage in the future life, and, as the authors point out, scenarios are possible in which a future child will gain an advantage through some kind of disability (Savulescu and Kahane 2009).

I believe that the principle of procreative beneficence fits perfectly in the context of space missions, where it obtains a much stronger justification than on Earth.

Environmental conditions justify the special concern of future parents for the condition of children born in a space colony. I make no moral distinction here between the application of health modifications and those of so-called enhancement, justifying their equivalence by the specific environmental conditions. Enhanced resistance to cosmic radiation and protection against rapid loss of bone density—assuming that only satisfactorily obtainable through GGE—are obvious starting points, but they are no less important than other potential modifications also obtainable through GGE that will adapt the individual to life in space colony conditions.

The only risk I see here is the risk of creating a more or less homogenized population that may result from applying similar or even the same GGE. One could add that concern for diversity is incompatible with a literal reading of the principle of procreative beneficence, which addresses the private relationship between a parent and their future child, not the vision of society of which that child will be a part in the future. What is good for the population as a whole might reduce the child's chances for future life, although I believe that the cosmic environment might in some situations reduce this loss of potential goodness to a greater extent than the terrestrial environment.

Perhaps the power in a space colony should have the right to regulate the reproductive freedom of the parents—because expecting self-regulation through voluntary decision may not work precisely because of living in a harsh, demanding environment—but any coercion by the power is the last thing humanity living in a space colony could desire. On the other hand, there is a concern that the specific population dynamics in a space colony cannot do without interference from authority at some stage, and perhaps what will suffice is procreative education, persuasion, and counseling.

The space environment, despite the aforementioned risk of homogeneity, on the other hand offers a special justification for the concept of procreative beneficence on the grounds that it offers relative environmental stability and thus may allow for better design and adaptation of desired features to predictable environmental requirements. Certainly this is the point of greatest controversy, which is a blight of opportunity but also a potential area of risk for the future population in space.

Whether on Earth or in space, even though the concept of procreative beneficence does not explicitly mention what traits are desirable for modification or selection, the concept of GGE assumes that we must make value judgments and indicate which traits we find desirable. This is true even for GGE for disease prevention, where, by the reverse route, we show what we value by indicating the disease we want to avoid (Baxter 2021).

Objections to the Procreative Beneficence Principle

The postulate of procreative beneficence grows out of the welfarist philosophy of wanting the best for our children. One may wonder whether the procreative beneficence principle is too strong. Its educational counterpart would be a principle that would dictate that future parents must guarantee their future children the best up-bringing and education (Blackford 2014, 15).

Another risk is the emergence of a race between parents and their modified children to continually increase the degree of a trait for the sake of others who also choose to benefit from biomodification. This occurs when enhancement becomes a positional good (Blackford 2014, 39–40). Nevertheless, enhancement is a good in itself for improving life, not a positional good whose purpose is to gain an advantage over others (Harris 2007, 29–30). The purpose of astronauts' enhancement is not for them to obtain positional goods that will make them superior to the rest of the population, but to guarantee their safety in order to successfully accomplish a mission. Sparrow accuses Harris and Savulescu of pursuing, perhaps unconsciously, historical eugenics. He also believes that the principle of procreative beneficence realized through mass GGE will lead to coercion of all prospective parents, who will not be able to fail to apply GGE under conditions of intense genetic competition in a society where such modifications will be massive (Sparrow 2011). This is therefore an indictment of the structure of society, not of the means that maintain it. As long as we live in a culture that demands competition and places a premium on competition, we will continue to seek out and use any means that can guarantee us a competitive advantage.

In the context of space exploration, we can envision a scenario in which a future society living from the beginning in a space colony would obtain certain improvements only, or primarily, through genome editing, be it SGE or GGE. Let us imagine that instead of many years of traditional education, future parents will be able to apply GGE, which will make it possible to achieve such intellectual and cognitive skills that children gain only after several or even more years of education. This does not have to be compulsory, nor the only form of education, there may be alternatives in the form of traditional education known on Earth, as long as the population in space is able to guarantee traditional education. On the other hand, children who will be subjected to GGE of an educational nature will be able to devote those few years to other tasks.

A Strong Case for GGE—John Harris and Robert Sparrow

Harris argues in favor of human enhancement in that if we can safely improve our physical and cognitive potential, then in principle we should do so. As Harris explains, every enhancement must be beneficial to the individual, because then it would not deserve to be called an enhancement (Harris 2007, 2011).

According to Sparrow, our obligations to future persons begin the moment we decide to bring them into existence. With that moment, we find ourselves in an ethical position of obligation not to harm that future person as early as their embryonic stage. Moreover, since in certain situations the failure to act is de facto no different from an intentional act in terms of its consequences, the failure to modify genetically at the embryonic stage can be interpreted as causing harm to a future person—however currently in the embryonic stage (Sparrow 2021).

The presence of the technique of human enhancement at the embryonic or pre-implantation stage may lead to a situation in which the general principle of beneficence, not only its specific form of procreative beneficence, may imply the obligation to use GGE at least in some cases—in those in which the principle of beneficence will make the use of GGE a necessary procedure. Thus, if I know that I can safely use GGE in a given case, and it is the only tool for guaranteeing the welfare of a future child, forgoing GGE in that case will be tantamount to a violation of the beneficence principle.

But, moreover, such an omission is also tantamount to a violation of the principle of nonmaleficence, in a situation in which the conscious decision to reject a safe and effective method of GGE will certainly lead to the deterioration of the future child's condition, and at least will not prevent it, despite the existence of an effective method of eliminating a given factor decreasing the quality of life of the future person or even the normal development of the embryo. Thus, as Sparrow rightly points out, the failure to use GGE where it is the only method guaranteed to increase the child's well-being, or at least to prevent a greater deficit or decrease in that well-being, leads to a violation of both the principle of beneficence and the principle of nonmaleficence (Sparrow 2021).

The space environment further justifies the application of GGE and allows one to consider it a moral evil, and at least negligent, for parents to opt out of applying GGE. Let us imagine a situation where, in a future space colony, parents can apply any kind of GGE that can exercise selected cognitive and behavioral functions. However, for some reason, some parents choose not to do so. As a consequence, their future child may face some difficulties that are not at all of the nature of a lack of competence in a competitive environment, but may be annoying only on a purely subjective level, where this future child, already as an adult, may regret not being able, for example, to control stress better or to control negative emotions in a given situation. Let us assume that the parents decided to bring up their child properly, guaranteeing her as much education, as well as moral and psychological upbringing, as was possible in the conditions of the space colony, but it turned out that better effects would be achieved by GGE.

In conclusion, the main argument for the moral justification of GGE comes from the principle of procreative beneficence. This principle is a refinement of the beneficence principle and states that parents have a duty to provide the best possible quality of life for their future children.

When GGE Can Be a Moral Obligation

An available GGE technology will offer complete safety can become a moral obligation for future parents especially when alternatives are not available (Battisti 2021). Among other things, Davide Battisti emphasizes the lack of availability of alternatives as a necessary condition. The question arises when, on the one hand, we agree that GGE may be the only option and, on the other hand, there is a risk that applying GGE will violate some principle such as precautionary principle, a duty not to harm nature, and a duty to future generations (Boylan and Brown 2001, 157). I believe that as long as GGE is the only option available, its application should not be prohibited, given the safety criteria. The application of GGE in situations where there is some risk should depend on whether the person subjected to GGE would be worse off than they would be without GGE.

Could GGE be our moral obligation in the space colony? Undoubtedly yes, if we consider it through the prism of preventive and therapeutic functions (Powell 2015). But there are strong reasons to believe that we should also apply human enhancement for nontherapeutic purposes as long as behavioral adaptations to harsh space conditions remain a challenge to currently applied countermeasures.

In sum, GGE can be considered a moral obligation when it is the only alternative. But even then, when alternatives exist, GGE may be justified when a child born with GGE is better off than they would have been without GGE.

Genetic Selection

Genetic selection involves making choices about which embryos can be born based on their genetic characteristics. The goal behind genetic selection is to have a child with a particular trait, or at least one who will be devoid of a particular trait that is considered negative. The problem arises at the moment when a trait is defined as undesirable or desirable.

Negative Selection

Negative selection refers to those interventions designed to prevent the appearance of a trait considered undesirable because of its expected harm. Negative selection is for prevention or treatment and is intended to eliminate undesirable traits from the population. According to Glannon, negative selection grows out of the principle of beneficence (Glannon 2001, 104). Glannon refers to such selection as negative eugenics (Glannon 2001, 102).

It is not difficult to imagine a scenario in which negative eugenics is legally sanctioned in a future space colony. Such a scenario is possible, however it would require several conditions that are not at all obvious. One is the existence of significant environmental pressures to minimize the existence of certain traits in the population in space. By this I mean a pressure that is the product of the significant influence of an adverse environmental factor and the limited capacity to respond to that factor when individuals susceptible to the trait emerge. Such a trait might be, for example, a lower degree of resistance to space radiation and any medical condition or disability that would make the individual require a commitment and infrastructure that would undermine the performance of the space base or colony.

We do not know how a future space base or colony will function in terms of number of members, social structure, and division of responsibilities. However, we can envision a scenario in which, at least in the initial stages of such a colony, each member will be required to work for the common good. But for such a scenario to occur, the trait undesirable in the environmental conditions of the space colony must be a genetic, heritable trait so that we can prevent its further transmission in the population.

Another condition is that this feature cannot be eliminated by any available means, such as drug and gene therapy. It seems, therefore, that the possibility of negative eugenics in a space colony is very small, and can be minimized even further if, bearing in mind in advance the risk of the emergence of conditions that might foster negative eugenics in space, we design life in the space colony in such a way as to provide an appropriate infrastructure and social structure so that the emergence of any "undesirable" traits from the point of view of a dangerous space environment will not pose a problem.

Positive Selection

Positive selection, referred to by Glannon as positive eugenics, refers to the improvement of a trait and function that is currently possessed by an individual to a normal, species-typical degree. According to Glannon, it is a kind of perfectionist improvement of the species, an effort to improve the population (Glannon 2001, 102).

The challenge of perfectionism (Roduit et al. 2015) poses no ethical problem for the application of human enhancement. This is not just because there are many different visions of human enhancement. It is primarily because this objection is inapplicable in the context of space missions. If we assume that the main purpose of applying human enhancements is to protect health, ensure safety, and increase the effectiveness and success of space missions, then the perfectionism criterion may lie in the background of decisions about the types and intensity of applied modifications, but it does not interfere with or hinder their application.

In the space context, perfectionism may simply mean applying the best, most effective biomedical agent currently available. Moreover, the principle of procreative beneficence does not presuppose or require recourse to perfectionism, but to what prospective parents believe can serve to increase well-being. Sometimes, this may include leaving an embryo disabled, which tends to conflict with our intuitive idea of what is perfect. Moreover, in the context of the cosmos, positive selection will be oriented toward mission goal and mission success, and this will be the guiding principle of genetic selection.

Finally, let us return to the term "positive eugenics." As in the United States in the first half of the twentieth century, a space colony might create conditions conducive to the evolution of a policy of positive eugenics, which in the United States consisted, among other things, of encouraging the reproduction of those individuals who had traits deemed desirable (Yashon and Cummings 2009, 235). Is such a scenario realistically possible in a space colony? Yes, but only if we identify traits that are genetically caused and inherited that are both strategic to the survival of the human species. And if so, positive selection understood in this way can be carried out still on Earth, deciding on the choice of candidates for a given mission, which depends on the nature of the mission and whether it is to be exclusive as today's space missions, or is to be more accessible to a wider group. Only then do I believe that positive eugenics may be acceptable, but the introduction of positive eugenics does not imply the introduction of negative eugenics, which would prohibit the reproduction of individuals with traits that are considered undesirable. The two types of eugenics are not correlated.

In summary, the concept of genetic selection is rooted in the principle of well-being and the principle of beneficence. It is difficult to find any moral constraints, such as principles of autonomy and respect for rights-holders, which could block the well-being principle driving the concept of genetic selection and thus stop the good consequences.

Mason's 500-Year Plan for Space Colonization

Mason proposed a vision of the next 500 years in which genetic modification will be applied essentially without restriction, primarily for space colonization (Mason 2021).[1] Mason refers to the human capacity for very long-term thinking about the species, and the resulting "extinction awareness." While this is on the one hand a common-sense and seemingly intuitive assumption, it conflicts with our carelessness about the welfare of the Earth and our destructive abilities. This awareness, according to Mason, but also many other scientists and philosophers, is to lead to the concept and, in principle, the necessity of colonizing the universe. From the risk studies perspective, one might question the concept of space refuge as the only salvation for saving humanity. Thus, the author's very underlying assumption can be questioned because people can equally well survive a cataclysm in shelters on Earth, and also these people can be appropriately genetically modified, adapting them to various new environmental challenges in a postcatastrophic earthly reality. But if we ignore this premise, we have strong reasons to agree with Mason's main idea expressed in the following words: "It is no longer a question of 'if' we can engineer life—only 'how.' [. . .] Engineering is humanity's innate duty, needed to ensure the survival of life."

Mason encourages us to use the potential of our knowledge and technology to engineer life. The purpose of this directed evolution is to enable us to settle in an inherently unfavorable and hostile space in order to protect our survival as a species. Mason discusses his research on the astronaut Scott Kelly genome from the NASA Twin Study. The main objective of this research is to identify the effects of the space environment on the human body for future long-duration missions, as well as to develop countermeasures. Mason points out that probably the biggest challenge for a flight to Mars will be strong radiation exposure. Mason suggests that the list of currently used physical, pharmacological, and medical preventive measures be enriched with an element not yet used in space, genetic engineering.

Mason discusses peculiarities, challenges, and targets of each of the ten phases into which he divides the space colonization project associated with the development and application of genetic engineering. The first phase planned for 2010–2020 is a detailed study of the human genome. The second phase, scheduled for 2020–2040, is dedicated to preliminary engineering of genomes. One promising example here is the creation of human cells containing the damage-suppressor protein present in tardigrades, which are known for their extraordinary resistance to radiation. Mason also discusses the possible usefulness of other techniques such as CRISPR. All genome editing methods aim to improve life on Earth, but they should also be developed and applied with future space travel in mind.

The next stage in Mason's project is long-term trials of human and cellular engineering, which falls between 2041 and 2100. He focuses here on preventative genomic defenses to counter the biggest threat on the journey to Mars, space radiation. This is the period in which "all the genes, cells, and even potentially organs of any organism can become a component of a human cell." The next phase 4 is preparing humans for space (2101–2150). As Mason says, it is a period of "cellular liberty" in which people will no longer be dependent on their genome. Mason expects that during this period genetic modification will be widely used, "and it is likely that a significant proportion of people in the United States will be zygotically edited or are the product from someone who is."

Phase 5 planned for 2151–2200 is the period of synthetic biology. During this period, people will be able to move fairly freely between Earth and space bases, and genetic modification will be available whether for specific jobs in various places in space, or simply for relocation, but also to improve vision in long-range missions where low-light vision may be desirable (Mason 2021, 154). Phase 6 (2201–2250) is about expanding the limits of life. Here Mason encourages us to open our minds to the idea that the possibility of life on different planets may require far-reaching engineering. Its purpose will be to make humans extremotolerant.

In the next phase 7 (2251–2350) people will live in a fully developed Martian colony, with a distinct culture including perhaps new types of religion. Intergenerational interstellar travels may be expected during this period. DNA of new life forms found in space could be sequenced, transmitted digitally, and then used to improve adaptations. Phase 8 (2351–2400) is focused on exoplanetary settlement, with phase 9 (2401–2500) "launching toward the second sun." Phase 9 is an era of genetically engineered optimism and nearly limitless human possibilities. Mason describes it as follows:

Humans have the ability to control their underlying genetic code, controlling for how their molecules fundamentally change in response to stimuli and enabling new abilities. This will enable an unprecedented ability to build, edit, and transplant cross-kingdom combinations of genomes, which we will need for the new worlds.

He rightly points out that the distinction between modification and therapy is likely to become blurred to the point where we are not so much talking about diseases as simply about our limitations, which we can overcome with genetic engineering. The last phase 10 (beyond 2500) describes a humanity who has fully possessed the ability to control her own evolution: "Humans will have a mastered an ability that truly sets us apart from other species—the ability to direct their own, and other organisms,' evolution."

Mason also adds that his idea is not just to survive the species or life itself, "but to achieve true cellular, molecular, and planetary liberty." Mason takes it for granted and necessary, justifying it by the duty to protect and extend life in the universe. However, life is not just about surviving, but about a quality of life. Will radically modified humans in a space colony be happy?

In sum, Mason's conception of a 500-year plan for space colonization grows out of optimism in the possibilities of science and is grounded in the assumption that from the knowledge we possess comes a moral obligation to apply it for the good of humanity.

Reproduction in Space

One can imagine two scenarios in which GGE and embryo selection for space missions can be considered. One is GGE applied on Earth for the purpose of selecting suitable candidates for future space missions, another is human reproduction in a space colony. Today, it is difficult to predict which scenario will be realized sooner, whether reproduction in space or selection by GGE of the best candidates for space exploration done on Earth. While the latter option is technically easier to implement, it may be more morally controversial even long after it exists as a technical possibility. Under the assumption that only adults will be sent into space, societies may object to modifying the embryo of future individuals who, at the earliest, will be sent into space after nearly twenty years of life on Earth or longer, if we maintain today's strategy of sending older astronauts into space. Such objections may indicate a reduction in effectiveness and performance during the years of life on Earth, and this is certainly medically possible, however dependent on the extent of the modifications applied and the adjustments expected. Thus, taking a more conservative approach to the political, social, and legal acceptance of such radical biomedical solutions as GGE for space colonization, it seems more realistic that GGE and embryo selection will be implemented first for reproduction in space.

Which of the following scenarios: the conservative approach (which opposes any reproductive selection), the approach which considers such selection acceptable, or the last approach which considers such selection not only morally permissible but even mandatory (Magni 2021), could be implemented in a future space colony? The first and third approaches, at least at first glance, are extreme approaches. Approach one is extreme because a total ban ignores the benefits that may accrue from the use of reproductive selection even if such a procedure guarantees safety and freedom from any negative consequences. As such, it should be rejected at least for a space colonization program or for space habitat. The space environment is too demanding to take as a default position a stance prohibiting all reproductive selection. It is possible that reproductive selection will be the only means of safe reproduction in space. It may not need to be used at a later stage in the life of the space colony, where the population of individuals born in space using reproductive selection will already possess the appropriate adaptive traits that will be passed on to the next generation during reproduction without the use of gene editing.

This is a biologically possible scenario, but probably questionable from the point of view of social and cultural development. Could a human colony in space really want to forbid the use of GGE just because it has already obtained a population with the required traits? It may want to do so from the point of view of population biology, whereas it will probably be so morally liberal that abandoning techniques that enable it to survive in space will probably not be dictated by moral conservatism, but by biological considerations. And if so, then we will obtain a population situation in which GGE is not needed for biological, pragmatic reasons. If we assume that a future population applying GGE in space must necessarily be permissive, such a population might abandon GGE purely for pragmatic, but not moral, reasons. Unless, again, it perceives a need to apply GGE for the purpose of obtaining some particular desirable trait that it cannot yet naturally obtain in the population. As an aside, one might add that a society that chooses to colonize space and is aware of these biological challenges and biological but also ethical consequences will be a rather liberal society.

This situation demonstrates two important assumptions exposed in this book, also indicated as central to space bioethics. The first is the conviction that the moral ecology of space is different from the moral ecology of Earth. Whether this is a qualitative or quantitative difference remains difficult to resolve, and some analogies to the terrestrial environment are legitimate, such as military ethics. However, this does not change the fact that at the starting point we obtain a very dangerous environment for humans, and this environmental element definitely influences bioethical conclusions. A challenging, demanding environment allows us to do more than we would otherwise be able to accept.

The second assumption is that the regularity inherent in space bioethics is its issue-driven and case-driven character, of which specification and particularism is an important methodological element. The case of reproduction in a human colony in space demonstrates the unhelpfulness or even harmfulness of the theory-driven approach, especially the deontological perspective. If we assume that at least two of the duty-based principles such as the principle of autonomy and the principle of respect for the person exclude GGE, then we can say that the deontological perspective or more broadly duty-based approach to ethics will hinder or even prevent full human expansion into space. Even if we assume that these duty-based principles should be implicitly applied as broadly as possible, it is possible that the space environment is the unique moral ecology that entitles us to abolish these principles at least as literally understood on Earth.

The issue-driven specification approach indicates that there are situations where the specific context is so unique that an otherwise valued and respected moral norm should be suspended. One can speak here of the influence of other principles that get the upper hand over duty-based principles, such consequence-maximizing principles as beneficence or nonmaleficence. But even this fluidity of transition between principles would not be possible with a classical approach to ethics based only on deontology or only on utilitarianism. In this case, utilitarianism may indeed hide under the discussed case study as a theory supporting reproductive selection, but there are bioethical situations in the cosmos where utilitarianism or consequence-maximizing principles more broadly should be suspended.[2]

The third approach to reproductive selection discussed, recognizing it as a moral obligation, is controversial. It stems from the belief that knowing that and knowing how implies moral obligations. At the very least, this position can be allowed for human reproduction in space on the condition that one does indeed recognize the positive effects of reproductive selection on population growth in space, and that other alternatives are not available, or at least not viable.

However, imagine a scenario in which only GGE for protection against cosmic radiation and the effects of altered gravity would be mandatory. Assume that some parents will object to a nontherapeutic GGE that, while not mandatory, will be recommended and will improve welfare and, in the context of a space colony, also increase safety by improving intelligence and/or morality.

Anomaly considers a similar case in the context of Earth, where unmodified children could use modern technology in ways that threaten the modified ones, which in turn could justify by the latter the imposition of either compulsory human enhancement or a ban on reproduction (Anomaly 2020, 89). Perhaps such coercion would be justified in the case of a space colony. We can imagine that a group of unenhanced individuals would be tempted to gain an advantage over the others by, for example, taking control of the life support system and blackmailing the others. It is possible, however, that the space colony will consist from the beginning only of individuals who will have the appropriate moral and cognitive predispositions, so coercion will not even be considered.

In conclusion, reproductive selection in a future colony in space sounds controversial at first glance. Nevertheless, a consequentialist perspective favors the consideration of reproductive selection due to environmental conditions.

Selecting the Best Candidates on Earth

Even if we imagine a scenario where GGE would be applied on Earth for future space missions to select the best candidates, this is a rather remote scenario. But let us assume that this will be the case when participation in space missions is associated with benefits to the participating individuals. Thus, the minimum condition for considering such a procedure is to pursue a space program that prefers or even requires candidates subject to GGE. What is questionable is whether an individual will want to participate in a space mission. The moral assessment depends on the effects of the applied modifications on life on Earth.

Another scenario is that continued life on Earth may be possible but will be a kind of constant, drawn-out global catastrophe. In this scenario, people will view participation in a space mission understood as a kind of space refuge as something attractive. Do parents have the right not to subject their future child to GGE? Do they have the right to deny them the chance for a better life? The moral assessment depends on whether it can be established with 100 percent certainty that this future child's life will be unbearable on Earth and incomparably better in space refuge. Perhaps there are scenarios in which life in a space refuge will always be unbearable to some degree, and perhaps it is better to live in unbearable conditions on the Earth we know than in a space we do not know and which threatens and terrifies us to some degree.

In summary, the concept of selecting candidates on Earth via GGE for future missions would require having foreknowledge of future living conditions on both Earth and space, and an infallible belief that determining a future child's life either only on Earth or only in space would be better for them than the alternative.

Sexual Selection and Social Engineering in Space

Sexual selection has been at work in our species probably since the very beginning of its existence (Gray and Garcia 2013). Partners decide who to have a relationship with and who to have a child with. People have been deciding for thousands of years with whom to reproduce in the expectation of passing on the best possible qualities to the future child. Often parents or other older members of society have made these decisions for their partners, usually the woman. Moreover, it seems quite reasonable to conclude that our entire culture has been harnessed to the needs of sexual selection (Miller 2000)—or is a product of it, including religion (Van Slyke and Szocik 2020).

What is significant about this practice is that the very idea of selection for reproduction has been practiced by humans since the beginning. Sexual attraction to people with certain characteristics such as that we prefer slimmer rather than obese partners, young rather than old, taller rather than shorter, perceived as beautiful or handsome rather than ugly, has its evolutionary origins. It is not the mere idea of selecting the characteristics of a future child that should be controversial. What can and should be controversial is the possibility of preserving reproductive rights and procreative autonomy. The moral problem will arise when, for environmental and/or population reasons, actors external to parents manage sexual selection. This is a scenario that could take place in a space colony for environmental reasons and limited space and resources.

Assume, however, that sexual selection in various forms will take place in a space colony. This issue also partially overlaps with the issue of parents deciding the fate of future children, making decisions about abortion or GGE. If such decisions must be made, should they be made only by the parents, or can such rights be granted to a third party based on the welfare of the colony?

We can imagine a kind of eugenic social policy in a space colony, such as Mars, where a third party would shape the reproductive rights of the colony's inhabitants. First, not everyone can have reproductive rights. Second, those who will have such rights may be forced or at least encouraged to reproduce. Third, migration from Earth may be limited only to appropriately selected candidates, where other criteria such as wealth or possession of enhancements will be considered alongside.

I believe that only a very strong rationale in favor of pursuing a space mission could morally justify a vision of society that encompasses these three points mentioned above. But even then, it is worth asking whether such a vision of society guarantees a life worth living, and whether it would not be better to respond in an antinatalist way that it is already better not to continue the existence of our species (because I assume that neither scientific nor commercial goals in space will ever justify such a vision of the population in space) than to agree to create, in a sense premeditatedly and with all consciousness, a society from the beginning that interferes so deeply with human freedom.[3]

Unless we are able to identify a scenario in which, after weighing the benefits (keeping the species alive) against the losses (restrictions on human freedoms), we decide that life is worth living even at the cost of such restrictions, which have a purely objective source in environmental specifics. Then the GGE ban, even if it were to be temporary and only cause a delay in the space mission, could in turn cause bystander suffering as a result of postponing the mission and not minimizing the suffering felt on Earth.

The task of futures studies is to outline also those scenarios that we want to avoid, in order to prevent early enough a sequence of events that, even against our will, may lead to a vision of a society—whether on Earth or in space—that will not be able to guarantee freedom and welfare. Nevertheless, I believe that, except in exceptional situations that we should prevent, applying GGE to a space mission is morally justified.

Green calls reproduction in space an ideology and a social experiment, which he compares to social experiments of the past. These two "experiments" cannot be compared because in space reproduction and the technologies applied to it are considered necessities, not trivial or extravagant experiments. Whereas on Earth it was indeed an experiment. Green also demonizes gene editing in space. He makes the mistake of saying that a simpler solution is to modify the environment. But how can we modify the microgravity environment other than through the use of expensive, complicated, and today unrealistic artificial gravity? Green also believes that gene editing for space missions will change humanity (Green 2021, 200). But such an argument is unfounded, because changing one or a few traits in one or a small number of people does not change humanity.

In sum, sexual selection and GGE in space are controversial solutions. Acceptance of them would require recognizing them as truly necessary for the survival of the species, but even then we might choose the antinatalist solution of saying that sometimes life may not be worth living.

5
Justification of Human Enhancement versus Rationale for Space Missions

Introduction

It should become clear by the last chapter that I am proposing an approach in which the moral justification for a biomedical procedure, in this case human enhancement in general and genome editing in particular, including GGE, depends on its context and purpose. This is, in fact, partly what bioethics is, which on the one hand is supposed to balance risks and potential benefits (which often leads to a conservative attitude, as in the case of GGE today), but on the other hand, it should also promote new biomedical technologies and emphasize their usefulness. In this chapter, I want to consider how the particular goals of space missions may affect the rationale for subjecting future astronauts to human enhancement primarily through gene editing.

As we saw at the end of the previous chapter, today an unacknowledged consequence of space colonization (which is often spoken of as the best, necessary, and only way to save our species) may be the need to establish a totalitarian society, which may call into question the sense of continuing the existence of our species under such totalitarian conditions.[1] As I have suggested, GGE may lead to such a vision, but it may just as well be a means that provides tools to facilitate its avoidance, at least so long as the very idea of space colonization must involve some vision of a totalitarian society.

Advocates of an absolutist approach to moral norms and principles in bioethics, especially advocates of deontologism, may not accept my position that links the moral acceptance (or prohibition) of biomodification for space missions to the justification for those missions. But the link follows logically, though not explicitly, from the research question posed by Harris in the introduction to his *Enhancing Evolution*. Harris asks:

> If the goal of enhanced intelligence, increased powers and capacities, and better health is something that we might strive to produce through education, including of course the more general health education of the community, why should we not produce these goals, if we can do so safely, through enhancement technologies or procedures? (Harris 2007, 2)

The question of the legitimacy of biomedical procedures applied for the sake of enabling participation in space missions can be framed as follows: if space missions will

The Bioethics of Space Exploration. Konrad Szocik, Oxford University Press. © Oxford University Press 2023.
DOI: 10.1093/oso/9780197628478.003.0005

enable us to live a better life or just to live (survive), and will bring some good to humanity or at least to the individuals who will participate in these missions (Baum 2016), then why not carry out these missions if it is a prerequisite to apply human enhancements to the individuals participating in it? Human enhancement becomes here only a tool, but a necessary one for achieving the good. The emphasis is not placed on human enhancement itself, but on the goals associated with the realization of space missions. The counterparts of the goals associated with human enhancement as posed by Harris are, in my view, the goals associated with space missions. If we agree that the goals expected to be achieved in the course of accomplishing these missions are goods, then we have no reason to regard human enhancements applied to their accomplishment as something morally wrong.

In the previous chapter we saw why we should apply GGE and embryo selection in at least some scenarios of our future in space. We were also able to discuss selected contexts that are considered controversial today on Earth. In this chapter I discuss the rationale for human enhancement primarily through genome editing with respect to three major types of space missions: scientific missions, commercial space exploitation, and missions conceived as a chance to save the human species.[2] I want to show how ethical and bioethical status of particular missions will be challenged by particular kind of rationale.

I do not analyze whether we have compelling reasons to pursue a particular type of mission.[3] I believe that over time we will be able to see the benefits of each of the mission types discussed, as well as have the resources to pursue them.[4] What interests me most is what the ethical status of human enhancement is for each type of mission, under different scenarios (ones that assume these missions are necessary as well as ones that question their necessity). My main hypothesis is that the ethical status of human enhancement depends on the type of mission being carried out and who may be the beneficiaries. But, as I show in the chapter, paradoxically, any type of mission justifies human enhancement as long as it is required by environmental conditions and human participation in that type of mission is necessary (that is, there is no automated mission alternative), and the principles of autonomy and beneficence are preserved.

In summary, the rationale for applying human enhancement depends on the circumstances, that is, the rationale for particular types of space missions.

Research Missions and Scientific Exploration of Space

The value of scientific exploration of space has been widely discussed by James S. J. Schwartz (2020) and Gonzalo Munévar (2023). For the purposes of my discussion, we need to discuss their arguments not only by assuming that science missions are valuable but also by considering such scenarios where perhaps they should give way to other purposes of space exploration. This, however, is not the subject of my interest. What is interesting and relevant for bioethical issues is the following question: if any human research space mission (to Mars and beyond, I assume here that lunar missions do not require human enhancement) requires application of human enhancement, then should we allow for such modification? Does value of science in space exploration justify human enhancement?

I agree that scientific purposes are not trivial. I also agree that scientific exploration as such is a high value in human axiological hierarchy.[5] However, we know that we cannot sacrifice everything that we have or who we are to make possible progress in scientific explanation. We do not accept tests on some animal species and we strongly criticize the infliction of suffering on those animals that, at least for now, science needs. We also usually agree that the moral status of human beings is higher than all other nonhuman animal species. If we are opposed to the use of animals for scientific research, perhaps we should be even more opposed to the use of humans.

Space scientists and philosophers argue for or against robotic or human missions. While some of them prefer automated missions,[6] some scientists emphasize advantages of human astronauts in the field.[7] This is ethically neutral as long as no human enhancement is required even for relatively long research missions. But if human enhancement is obligatory even for short research missions, then are not we treating the modified astronaut a bit like a lab animal?[8] This becomes clearer in the context of the dispute between supporters of human and robotic missions. We can assume that due to the high risk and high costs of space missions, automated missions should be treated as a default mode of operation, at least to conduct research in space. We can also assume that humans are sent to make research in space only when really needed, when none of the currently available robots can perform the planned research as well as a human. This is an instrumentalization of the human astronaut that is not bad or, at least, morally challenging, as long as no enhancement is required and no unacceptable risk to astronaut health or life is expected. But the ethical issue arises if scientific purposes of space exploration may justify radical human enhancement of astronauts. If yes, we should ask what kind of scientific objectives are so valuable (and why).

The main goal of scientific exploration of Mars is to investigate whether life once existed on Mars, as well as whether, even if life did not exist, Mars could be habitable (Des Marais et al. 2008, 599). Another associated goal is to explain planetary evolution (including formation of the planet, and its climatic and biologic history) and compare it to the evolutionary history of other planets, including Earth (Carr 1996, 185–186). Obtaining definitive answers about extraterrestrial life, either confirming or ruling out its existence, would have important implications for culture, especially for our knowledge.[9] As Steven J. Dick suggests, it could be that as a result of discovering the existence of extraterrestrial life, we would look at our knowledge generated on Earth as merely part of a larger whole, as part of a larger perspective (Dick 1998, 273).[10]

It is worth pointing out a specific problem related to justifying the sense and need for scientific space exploration. Let us assume that we are skeptical about the project of a scientific space mission. But we do not know whether the knowledge gained in the near future in space exploration will serve, perhaps, in the somewhat distant future to make significant progress in some important field of an applicable nature. It is worth remembering that many purely speculative discoveries, seemingly useless, have made possible important practical changes (Flexner 2017). Thus, basic research from this point of view makes sense even in terms of its potential future applicability.

Munévar points to two other qualities unique to space research. First, since it is inherently global research, it enables a better approach to and solution of global problems of humanity, such as climate change in the first place. Thus, the perspective inherent in space research enables the adoption of such centralized optics. Second, and perhaps more metaphysical and ethical than practical—but which may lead to practical consequences—is Munévar's observation that learning about other worlds will give us a better understanding of our world. Finally, and particularly important for countering social and ideological criticisms of space missions, Munévar points to a number of material and social benefits of space missions (Munévar 2023).[11]

What I am interested in is not whether space science missions are really in themselves essential to the survival of our species, and at least to improving its wellbeing, and whether they are a kind of extravagance. What I am interested in is whether we should accept the radical human enhancement of future astronauts—assuming for the sake of argument that the presence of humans in the field is essential, and that due to the length of the mission and the distance from Earth it is necessary to apply genome editing—sent solely for the purpose of carrying out scientific missions depending on the importance assigned to it, practical or theoretical.

Imagine being able to be assured that a space mission will produce knowledge of strategic importance for the survival of humanity, for example, to combat climate change. Subjecting future astronaut-scientists to genome editing is not morally controversial. What might be challenging is the risk of side effects and readaptation to Earth conditions, if such a problem could realistically occur with this type of modification. But would not such a mission be like a military mission, where the survival of an entire nation is at stake? Let us assume that in addition to professional astronauts, who would accept the need to apply genome editing as part of their training and necessary preparation, we need to recruit a group of at least a few scientists not previously involved in astronautics. In addition to astronaut training, they must be exposed to genome editing. The space agency calls for applications from volunteer scientists, including in the list of requirements their agreement to undergo mandatory genome editing. I believe that even with genome editing still controversial, once the scientific world is assured that a scientific space mission will yield knowledge of revolutionary importance to the survival of the species, there will probably be a few among the large population of scientists who will agree to undergo genome editing.

However, if no volunteers are obtained (assuming that the obstacle is the obligation to undergo genome editing rather than the hardship and risk of the space mission itself), I object to forcing any scientist to participate in such a mission on the grounds that the principles of autonomy and respect for the person must be respected. This is the case where I reject utilitarianism, not allowing the sacrifice of the individual for the larger population. The alternative then remains the postponement of the mission (since the condition of the proposed thought experiment is the assumption that human presence is necessary, which rules out teleoperations, videoconferencing with scientists from Earth, and intelligent robots) and, for example, an appropriate education campaign.

Because the scenario discussed above describes the situation with the highest possible degree of justification, each subsequent scenario will strengthen resistance to forced genome editing. In contrast, I do not oppose voluntarily applied genome editing even for such scientific missions that are not guaranteed to produce knowledge that is relevant either practically or theoretically. I believe that if the conditions of safety and efficacy are met, anyone taking part in a space mission with scientific purposes can voluntarily accept to participate in such a mission and to undergo genome editing.

In conclusion, the application of human enhancement for scientific missions, especially those of a significance deemed important to humanity, is morally permissible as long as the rule of informed consent is upheld.

Commercial Exploitation of Space (Space Mining and Space Tourism)

The concept of space mining is a broadly discussed issue. Many works discuss technological peculiarities, economic expectations, and political context, some convincingly arguing for the many benefits associated with space mining, particularly asteroid mining, but also with broadly understood space industry (Pilchman 2015), (Metzger 2016), (Keszthelyi et al. 2017), (Pelton 2017), (Knowledgemotion 2018), (Paladini 2019), (Sivolella 2019), (Elvis 2021).

We can find here at least two different approaches or phases. One of them is a current or near-term phase when humanity lives only on Earth, and human activity in space is a kind of human extravagancy—even if useful, not necessary and not crucial for our survival. This stage includes not only space mining but also space tourism with orbital flights and space hotels. The second stage is a phase when humanity started space settlement and the issue of in situ resources utilization becomes a matter of survival.

An important emerging topic, which is probably more challenging in the first stage of resources exploration, is the legal status of space resources, and rights and duties of all sides involved in space exploration including public and private agencies. The issue of sustainable development emerges here together with the concept of global justice. While all these issues have their own high importance, they are not necessarily relevant to bioethical concerns. I want to consider the bioethical status of modifying humans for resource exploitation. The status of human enhancement may be different for the first and the second phase. I do not consider the third scenario where people live only in space settlement. I assume that in this scenario, exploitation of space resources is self-explanatory.

Phase 1—Commercial Exploitation of Space When People Still Live Only on Earth

There are two types of argumentation available here. One of them refers to the human right to conduct commercial activities. In this case, no special ethical argumentation is provided. The second case is definitely more challenging and more interesting ethically. It consists in the fact that the proponents of exploration draw attention either to the possibility of enriching the whole of humanity and simply improving global welfare, or point to the risk of depleting resources on Earth and the only rescue taking the form of space exploration.

Schwartz offers convincing arguments debunking this unjustified optimism of advocates of space mining. First, he shows that while the total virtual amount of space resources as such may be limitless, those resources that are available to human technology in the coming decades, and would be economically rational, are very limited. It is worth quoting a passage from Schwartz's book, which accurately expresses the absurdity of the belief that the exploitation and importation from space to Earth of certain primary raw materials would ever be profitable:

> While terrestrial supplies of nickel and PGM may run so thin as to make asteroid sources competitive in terrestrial markets, it would be absurd to say that humanity as a whole would benefit from, e.g., being supplied with water or aluminum from space. For resources like water or aluminum to become cheaper to buy from extraterrestrial sources would imply that the situation on Earth had deteriorated beyond a reasonable point. That is, if it is cheaper for terrestrial humans to purchase their water from space, then they are already doomed. (Schwartz 2020, 174)

Second, Schwartz does not expect that the beneficiaries of the exploitation of space resources may be those most in need, but rather—according to the specifics of capitalism—those already wealthy and directly involved in their exploitation (Schwartz 2020).

Let us assume, following Schwartz, that commercial exploitation of space in phase 1, that is, solely for capitalist purposes, not aimed at the survival of humanity and improving the living conditions of the Earth's population, is something we should view negatively rather than positively morally.[12] Theoretically, we could look at this phase 1 space exploitation project in a neutral way similar to how we treat resource exploitation on Earth, adding a correction for the risk of superexploitation, taking into account population ethics and sustainable development. However, it is impossible to ignore the criticisms of philosophers and ethicists of space mission who point to important issues of environmental ethics, the value of space itself, concern for its integrity, the moral status of inhabited and uninhabited space, and the risk of contamination.[13] These very serious warnings by philosophers undoubtedly diminish the moral legitimacy of the concept of space mining in Phase 1.

What would be the ethical status of the concept of human enhancement by genome editing of future participants in exploitation missions when their presence is required (assume that, at least in some tasks, humans cannot be replaced by robots) and genome editing is necessary?[14] As long as participation in such a mission, which would be treated as paid work, is voluntary and, moreover, likely to be financially attractive, the decision to undergo genome editing should remain within the scope of informed consent,[15] the autonomy and freedom of the individual interested in undertaking the work in question.

I omit here the ethical doubt of undertaking commercial space mining due to the environmental space ethics caveat. I believe that the obligation to undergo human enhancement should be treated as a criterion for taking the job required by the employer, indistinguishable ethically from other criteria. I make no distinction between therapeutic enhancement and enhancement directed at increasing job performance. The main argument against human enhancement is the issue of safety. Nor do I see the risk of social objections suggesting that modified individuals might gain an advantage in space or on Earth. As long as commercial exploitation of space does not involve the creation of a permanent base or habitat, living conditions in space will not generate a hierarchical structure analogous to Earth societies.

As for the risks of returning such modified workers to Earth, on the other hand, I believe that their gaining a possible advantage over unmodified people will not be problematic for at least three reasons. First, modifications for space may not be applicable on Earth. Second, even if they did, the number of workers working in space might be very small when compared to the population on Earth, which would not skew the labor market to the disadvantage of the unmodified. Third, even if the risks described in point two were to change, it may be acceptable to use safe biomedical interventions that deactivate adaptation for space missions.

Space Tourism

There are a lot of moral objections against engaging in space tourism that appeal to a misallocation of time, energy, and resources.[16] This is an objection often made in public spaces (CNN 2021). Nevertheless, neither the growing interest in space tourism nor its hypothetical importance as one of the vital links in the sequence leading to space colonization can be ignored.[17]

Perhaps the potential application of human enhancement for space tourism will be justified on health grounds even more than for other types of space missions, if only because future space tourists will not be professional astronauts or even pilots. They will therefore be people without proper training and perhaps of average health level.[18] It is this interesting aspect of space tourism that, in my opinion, opens up the possibility of applying human enhancement despite the fact that, paradoxically, the justification for tourist missions is the weakest from an ethical point of view (see accusations such as extravagance, wasting money, polluting the environment, and so on).

But at the same time, it is in these types of missions that we can expect to find the least prepared and physically and medically predisposed individuals who may require human enhancement more than participants in scientific or commercial missions other than space tourism. Thus, because of the specific profile of participants in travel missions and the most inclusive selection criteria, human enhancement makes the most medical sense for just this type of mission. The justification increases with the distance and duration of the mission even if it remains "only" a mission of a tourist nature.

Does the application of human enhancement for touristic missions raise any moral objections? I think not.

First, it is the most voluntary of all possible types of space missions, because there is no motivation in the form of financial benefit, which may in a way affect the autonomy of choice in the case of space mining. Thus, it is the mission type that guarantees the most autonomous choice.

Second, due to the high cost of participating in such a mission, there is minimal risk of negative social consequences of modified individuals appearing in the Earth's population to gain an advantage over unmodified individuals. In addition to the fact that it is unknown whether the modifications, which in the case of space tourism are likely to be purely therapeutic (as opposed to space mining, which may require or recommend modifications of an enhancement nature to increase performance and productivity), might cause any unfair competition effect at all, an additional factor is the age group that dominates today's space tourists, which is unlikely to participate in the market rat race. It will also be a small group that, even if it were to obtain unique adaptations also efficient on Earth for competition, will be too small to cause economies of scale.

Finally, third, human enhancement for space tourism may be recommended because of the long-term health effects on the aforementioned age group. It may be that the applied biomodifications will increase the overall immunity and performance of the body, positively affecting health. Consequently, at least some of the elderly will receive a special form of therapy that they would not be able to obtain without participating in space tourism missions. I assume, for the sake of the argument, that any human enhancement will be applied exceptionally only for space missions, and even a rich enough person will not be able to obtain (buy) human enhancement outside the context of space missions (as far as we can imagine such restrictions in free market conditions).[19]

This may lead to a somewhat paradoxical situation where wealthy people will be recommended to take part in space missions just to get a human enhancement, which for example will extend their life by a few years or simply have a positive effect on their health. This practice would not increase inequality, because anyway the financial and health status of a rich space tourist is consistently better than the statistical resident of both the United States and many other countries. And so this poor fellow would never receive human enhancement under the discussed conditions, because she would not be able to cover the costs of space flight or staying in an orbital hotel on Earth, the Moon, or Mars. Therefore, denying the right of the richest to obtain human enhancement while carrying out space tourism is a form of discrimination and a waste of resources that exist, and which in any case, for other reasons, cannot be distributed otherwise than to the richest taking part in space missions. In this sense, the practice adds nothing to the already existing inequalities in the distribution of financial and medical resources, because human enhancement is not a resource intended to be shared, and is therefore not part of public health.

Phase 2—Commercial Exploitation of Space When People Settle Space

This scenario not only strengthens the rationale for human enhancement acquired voluntarily by the individuals involved, but opens up possibilities for discussing a form of coercion as long as the exploitation of space resources requiring the participation in the field of modified workers is necessary. I believe, however, that it is possible to avoid introducing into the discussion a scenario that carries the risk of coercion to submit to enhancement for practical reasons. Since the described scenario requires considerable technological advancement, it can be assumed that enhancement will be applied much earlier for missions in earlier phases of space exploration, such as science or phase 1 exploitation missions. Therefore, the problem of coercion introduced into the discussion here appears to be an apparent problem.

It can be assumed that in the population living in space there will be a group of people who will voluntarily accept the obligation to undergo modifications. Perhaps this will be the group participating in the voluntary exploitation of phase 1. Moreover, there is a strong rationale for concluding that every inhabitant of such a habitat in space will be subject to compulsory modification for health reasons.

To sum up, the application of human enhancement for commercial space missions is devoid of situations assuming compulsion to undergo human enhancement, which makes such modifications morally neutral from the point of view of the individual concerned.

Space Refuge and Space Settlement

One of the main reasons for considering the idea of space missions as space refuge and space colony[20] is the threat of existential disaster.[21] One of its potential causes is the impact of a comet or asteroid. Even if we assume that the risk of an asteroid or comet colliding with Earth is extremely low, what is important is to determine whether our generation should take some action to mitigate that risk for the future, even if we have a warning time of decades. A near-Earth asteroid impact with the potential to destroy civilization, that is at least 2 kilometers in diameter, happens once every 2,000,000 years. As monitored by the Spaceguard Survey, they pose no threat for the foreseeable future. Comets, on the other hand, can pose a threat; they are difficult to detect only a few to several months before they enter the Solar System (Morrison et al. 2004, 354–355, 378).

A lot of ethical issues[22] may be discussed around this topic (Schwartz et al. 2021). We can question two basic assumptions at the foundations of the concept of space refuge. One of them is the belief that survival of human species is worthy of protection or even should be protected. The belief that humans as a species should be protected comes in degrees. One may believe that humans should be protected by standard means and routines. Routines include here healthcare system, police, fire brigade, and other social services and institutions that protect human lives. However, someone else may believe that we do not have to be, or we do not deserve to be, protected by extraordinary means. Extraordinary means include such situations as when a newborn baby has a very rare disease and only very expensive and highly complicated surgery can save her. We usually try to do that, at least in richer countries, but such a strategy could reach a critical level when so many damaged fetuses are born that there is a shortage of medical supplies. The same applies for euthanasia and assisted dying (Szocik and Abylkasymova 2022a). Some philosophers argue that humans simply do not deserve such extraordinary protection due to their immoral behaviors and destructive nature, which can be harmful to new objects in space.[23]

The second belief that may be contested is the idea that only space refuge may protect human species against certain types of disasters. There are good reasons to assume that terrestrial refuges will work as well. Anyway, what will be the difference between postcatastrophic Earth and the Martian or lunar landscape? In all places, life without life support systems will not be possible, and more to the detriment of the space refuge—even on a postcatastrophic Earth life without a life support system may be possible, unlike the space refuge. If we take into consideration such a perspective and ask what is next, what our life in the space colony will be like, we can find that it will be a kind of hard and stressful life, in a constantly high-risk environment.[24]

There are two things to keep in mind here. First, life in a space refuge may be even more challenging and risky than life of survivors on a postcatastrophic Earth (Kovic 2021). It is enough to note that Earth possesses parameters such as gravity and exposure to space radiation more attractive for us than other objects in space which we are able to settle.

Second, space settlement should not become our first and main target. Humanity should try to survive on Earth as long as possible independently in some way—even in subterranean and/or aquatic refuges—and independently of, perhaps developing in parallel, a program of space exploration and settlement.

However, the hypothetical advantage of shelter in space should not be forgotten in this context. Assuming that the latter will be reasonably stable and thus predictable, and that no threats will occur at the space base, the advantage of the space habitat will be the stability just mentioned. Its opposite may be a rapid change in the living conditions of the survivors of a cataclysmic event on Earth, problems with supplies, difficulty in providing basic needs, and thus unpredictability. This will be a collapse of civilization that will not occur in space.[25] A specific further advantage of a space base in this context is that a catastrophe in space is likely to lead to the rapid annihilation of all (although I realize that this is the kind of consolation that satisfies perhaps only antinatalists)—probably through the destruction of the central life support system. Thus, there will not be a handful of survivors concerned about their survival under harsh, postcatastrophic conditions in a space base.

However, if we take for granted two mentioned above premises, that we are worthy of protection by extraordinary means and that space refuge is such obligatory means—which, as we can see, are not obvious—we meet a couple of bioethical issues. Some of them are especially relevant for the idea of human enhancement.

Schwartz (2020) critically evaluates the basic argument underlying the concept of space settlement—our duty to extend human life. He argues that while space settlement as such is required for our survival in the long run, it is not required in the near future and should give way to alternative methods of enhancing humanity's chances for survival regarding purely earthly policies.

However, we can, for the sake of argument, assume that, at least in the long term, humans will be forced to leave the planet and find another habitat, even if that might be many thousands of years from now. We are talking about a period of time incomparably more distant than the timeframe for the other two basic mission types considered above. It is this distant time frame that proponents of making humans a multiplanetary species usually have in mind. With this approach, seeking and developing shelters on Earth is not a counterargument. If we assume that sooner or later all humanity will die on Earth, settling somewhere else becomes inevitable. And here are the first ethical difficulties for this concept, which are also relevant to the concept of human enhancement. I assume that human enhancement is contemplated as necessary for the accomplishment of permanent space missions, and that the motivation for its application in the form of saving the existence of the human species gives it a strong, arguably the strongest, justification (assuming that radical human enhancement still retains its ethical controversy).

The problem is as follows. It is not at all obvious that we have an obligation to save humanity in this way, as a species. Perhaps it is a belief that arises intuitively. Upon closer examination, however, there is some reason to believe that we do not have a strong case for the concept of space settlement dictated by the preservation of the existence of our species.

One counterargument is that space colonization is an extraordinary measure. It is not a standard procedure for preserving humanity and even individual lives. It does not follow from the reference to biological evolution that any organism, no matter how highly evolved and how long lived, has a duty to exist. Great extinctions have occurred in the history of life on Earth. From this point of view, it is natural not to prevent them. We also do not know what will be the conditions of life in today's even unimaginable place in the universe, which may be chosen in the distant future as the site of a space colony. For various reasons it may be a place unfavorable to life, and its future inhabitants, especially those who will be born there, may not want to be forced to live in such a place.

Here we come to a key point that should be discussed in the context of the space refuge concept, namely the rights and interests of future humans, and also our relationship to them. If we assume that it is our moral duty—derived from the principles of beneficence and nonmaleficence—to provide future generations with living conditions at least as good as our own, then colonizing a space object where living conditions would be characterized by a low level of quality would constitute a betrayal of these principles.[26]

Finally, our relationship to future generations is unclear. The relation to the next generation looks different, and the relation to virtual generations distant in time from us looks different. In this context, the arguments of the concept of antinatalism have an exceptionally high degree of rationality.

Alternatives to sending humans to distant places in space under the concept of space refuge are numerous. One of these is the concept of embryo space colonization (Edwards 2021). Perhaps somewhat paradoxically, this concept seems to be even more morally controversial because of the sending of embryos alone, and thus, after they develop from fetuses into children who will be deprived of care other than that of, for example, robots and AI. An alternative might be to emulate the brain and send something like microchips that have human consciousness.

The concept of space refuge and space settlement raises long-term ethical dilemmas. While, like all space bioethics, it is highly dependent on context and technological possibilities, the controversial nature of many of the bioethical issues surrounding space exploration is hugely dependent on the technologies. Perhaps radical human enhancement will always remain controversial regardless of the development of technologies that guarantee the safety and reversibility of the modification applied. But the space refuge case discussed here already loses its controversiality if the technology of the future makes it possible to guarantee a quality of life at least as good as that of the inhabitants of Earth. Perhaps some form of psychological bioenhancement would be not only recommended, but mandatory, assuming that today's "pharmacological optimism" (or genetic optimism about the possibility of psychological and moral enhancement) finds a real basis in the future.

Regardless of the resolution of the question of whether humans should make extraordinary efforts to prolong the existence of the human species through space colonization, or pursue alternative actions, or abandon any efforts at all, I assume that the claim that it is our duty to do so is not obvious. I therefore disagree with Mason's belief that such a duty is self-justifying, a kind of intrinsic duty: "Most duties in life are chosen, yet there is one that is not. "Extinction awareness"—and the need to avoid extinction—is the only duty that is activated the moment it is understood" (Mason 2021, xi–xii).

Mason's claim of a duty to save the existence not only of the human species but also of life in general, referring to the human unique capacity for self-awareness and consciousness of annihilation, also seems to conflict with the idea of biological evolution. From an evolutionary perspective, humanity is not unique in any sense of progression in evolution, nor can it be seen as the culmination of evolution. Moreover, evolution may have unfolded in a different way, with nothing to indicate the emergence of such a being as a human being. Finally, these distinctive characteristics of humanity concerning its intelligence and capacity for consciousness must be regarded simply as mere adaptations to environmental conditions, one of many that have evolved evolutionarily. From this evolutionary point of view, human intelligence is an adaptation to certain environmental conditions, but it is certainly not the culmination of evolution (Futuyma 2006, 519).

Assume, however, that there are good reasons for humanity's decision to need to colonize space to save humanity. Let us make several assumptions such as that it is necessary, that we will be able to organize such a project logistically and temporally (the challenge is to recognize a good time to begin the sequence of events leading ultimately to the building of a colony in space—a serious and costly risk is not only to take action too late but also too early),[27] that we reject antinatalist arguments questioning the wisdom of trying to prolong the existence of our species because of the expected increase in the sum of suffering in space, and that, finally, human enhancement will be necessary. Under these assumptions, human enhancement acquires moral justification because it is only one of the means—and not at all an extraordinary one—undertaken to realize the most important of values, that of defending and prolonging the existence of the species. The prominence of the unusual and extraordinary nature of even the most radical human enhancement diminishes when it is confronted with the plan for space colonization, which is definitely an extraordinary measure taken to protect the existence of the human species and perhaps other species as well, which strengthens the moral justification for human enhancement.

Since in this scenario human enhancement is necessary to enable evacuation from Earth to a safe place in space, any argument against even genome editing will lead to the death of either the individual who opts out of evacuation because of the need to adopt human enhancement, or of humanity as a whole if potential mission organizers were to abandon the space mission for this very reason. The equal access necessity argument proposed by Schwartz (2020)[28] loses its point here as long as, at the point at which leaving Earth is necessary to save lives to ensure the functioning of the space refuge, accomplishing such a mission without applying enhancement is medically impossible. It will not always be possible to delay the timing of a space mission in hopes of producing technology that eliminates the need for human enhancement. In the considered scenario, the alternative to human enhancement allowing participation in the mission is death on Earth. To convince individuals skeptical of the genome editing needed for life in space, an information campaign and perhaps an educational program would be needed.

I believe, however, that a much bigger ethical problem than the obligatory human enhancement generated by the concept of space refuge is the selection of candidates for space missions. A possible scenario is both a situation in which the right to participate in the mission is denied to some of the volunteers for various reasons, and a situation in which there are fewer volunteers than the number of people deemed necessary for the future operation of the space base. However, I assumed—because this is one of the conditions for this thought experiment—that the implementation of the space refuge would be possible for logistical reasons, including the challenge of candidate selection.[29]

In conclusion, human enhancement for space colonization is justified by the desire to protect the survival of the human species. A greater ethical challenge is posed by issues related to the concept of space colonization, such as our obligation to care for future generations and the very meaningfulness of space colonization, which is understood as a concern for humanity at all costs.

6

Is the Bioethics of Space Missions Different from Bioethics on Earth?

Introduction

Let us begin by considering to what extent the selected examples from human history on Earth may be analogous to what we may face in future long-term missions. I am thinking not only of Antarctic expeditions, but especially of journeys across the ocean, where explorers sailed for many months to present-day Australia, New Zealand, and North America. In some ways the ship environments are similar. Small space, no contact with land, travel into the unknown, inability to get help from land/earth, as well as inability to return in an emergency. The situation of the explorers of new lands on Earth was even worse than the astronauts because of the lack of communication. Astronauts in the worst case can only talk about the delay in the transmission of signals, which in the case of the distance between Mars and Earth will be a few tens of minutes at most.[1] Can we say that ocean travelers could do more, could we allow them to do more than people on land?

It is difficult to answer this question using today's ethical criteria, because of the moral progress that has been made from then until now. Throughout human history we have had similar phenomena in the sense of isolation and the impossibility of return, which, to some extent at least, can affect the moral qualification of a given environment.

In this chapter, I attempt to answer the question posed about the potential uniqueness of space mission bioethics. I mostly refer to the issue of human enhancement, although in a few places, especially comparing space bioethics with military ethics, I discuss issues such as autonomy and informed consent. However, I avoid considering issues that are currently known and debated, especially protocols for clinical research with astronauts, focusing instead on the bioethics of future long-term missions requiring human enhancement applications.[2]

Space Environment as a New Moral Ecology

I believe that the space environment is a new moral ecology, which implies the need to apply existing moral principles and rules to new situations that will allow flexibility in dealing with a new, unfamiliar, and dangerous environment.[3,4] Already, orbital flights

The Bioethics of Space Exploration. Konrad Szocik, Oxford University Press. © Oxford University Press 2023.
DOI: 10.1093/oso/9780197628478.003.0006

are making significant changes in such rules and concerns as privacy, the conduct of medical research, and clinical medical decision-making (Robinson 2004). What characterizes space mission bioethics and links it to extreme environment bioethics and military ethics is that the tolerance for the suspension of certain principles and rules is greater than in traditional bioethics.

I assume that the specificity of the space environment, as well as the exceptionally strong rationale for pursuing space missions make the concept of human enhancement take on a specific meaning in space. One difference between the concept of human enhancement for space missions and the way the term has been used by at least some authors is that, with respect to space missions, enhancement is not seen as a philosophical program to create a new and better human being. And that is how the concept of human enhancement is often viewed, in terms of a program to create a better human being, an overall enhancement of the human species (Hauskeller 2013).

Many philosophers and bioethicists do not discuss the concept of human enhancement in the terms of transhumanism. Not everyone who talks about human enhancement has in mind the creation of the so-called new human, the posthuman. Nevertheless, many philosophers refer to the concept of transhumanism and introduce the term "posthuman." The following definition of transhumanist modification is worth quoting: "Transhumanist enhancement: any intervention, not necessarily medical, aimed primarily at the improvement of one or more core capacities of an individual beyond species-typical limits with the aim of overcoming human biological limitations" (Cabrera 2015, 64).[5]

Many philosophers also, already independently of the use of these terms, often speak of such supermodifications as modification of intelligence and morality. These are modifications that, for the time being, are not possible from the point of view of science and it is not known if they ever will be. They are not modifications of the values of one small parameter, which can be regarded as one of many components of either intelligence, such as improving memory or concentration in learning, and morality, such as suppressing urges or putting one into a state of euphoria.

When philosophers speak of modifying intelligence and morality with gene editing, for example, they usually mean comprehensively raising intelligence or morality to a new and higher level. However, what characterizes space bioethics is the abandonment of both the use of terms such as "transhumanism" and the concept of the posthuman—it is not clear who such a posthuman would be in the context of space missions, and the abandonment of considerations of giving humans in space super qualities such as an almost mystical modification of intellectual and moral potential.

Although in the next chapter I discuss the concept of moral bioenhancement for space, I do not consider it as one of the main goals of possible modification of future astronauts and space colonizers, but rather I intend to discuss this concept in relation to the discussion on this topic currently taking place in relation to the terrestrial environment. By doing so, I want to show that if we were indeed ever to modify human morality—however we understand this very broad and vague concept—it is the space environment that might be the optimal environment.

However, with regard to the concept "posthuman" for space missions, it can be added that the only sensible use of a similar term would be in the situation of the speciation of a human population living in a space colony.[6] But it is not clear that such speciation would be possible when we consider that a human population in space could apply human enhancements quite widely and intensively. Such enhancements in addition to adaptive functions could also play the role of mutation-correcting factor and guaranteeing the preservation of the same genome, except for better resistance to selected factors of the space environment.[7]

The basic understanding of human enhancement is the one that points to such modifications that do not have therapeutic goals involving prevention and treatment (Gyngell and Selgelid 2016, 112). The cosmic context challenges this definition, which emphasizes the lack of connection of the applied modification to a concern for good health and excludes medical functions. It is from this that the distinction between enhancement and therapy arises.

How do we understand enhancement in the cosmic context applied to exceptionally healthy and fit people? Moreover, it is not entirely clear whether a particular modification will actually be necessary or merely a recommended factor in increasing the chances of maintaining good health during space missions. Perhaps some enhancements planned for application in space will have an effect similar to the Covid-19 vaccine applied to young, healthy people with strong immunity. Perhaps such people would become infected with the virus without the vaccine, but still the course of the disease would be slight, perhaps even unconscious to them. But perhaps as a result of the infection such an unvaccinated young person would suffer a serious collapse of health, and perhaps even die. Some modifications in space may work similarly, and perhaps the genomes of individuals will be variously adapted to respond to particular harmful agents in space, but only to a certain extent.

It is possible to consider this context of applying human enhancement for space missions as a kind of preventive action, but it is certainly not classical disease prevention understood in the sense ascribed to typical therapeutic actions. It follows that a major conceptual challenge to the concept of human enhancement in space is not the difficulty of defining disease states and establishing a so-called norm. These issues are discussed in the context of Earth. What presents an additional challenge is the application of enhancements, which are guided by therapeutic goals, to healthy people who may never experience any negative effects from exposure to particular harmful agents in space, and at least these effects will not require the application of radical enhancements.[8]

What further complicates this conceptual difficulty is that the modified astronauts will be beyond the norm inherent in the human species. Assume that for a long-term mission to Mars, astronauts will be modified to increase their resistance to cosmic radiation. While the motivation for such modification is medical (therapeutic, preventive), the effect is to produce a unique property.[9] By the same token, it is not the so-called catch-up that is implied by a therapeutic modification, applied to sick people or those who carry a certain gene that favors a disease or inevitably causes it at a certain stage of life. As a consequence of such enhancement, we get a very healthy and fit human being equipped with a super property that no one else on Earth has. Perhaps this property will not be needed, or its usefulness can be compared to the mentioned vaccine.

Some usefulness for the context of space missions is the definition of enhancement understood as the application of a change beyond a species-typical trait (Savulescu et al. 2011). If we accept this definition, we can consider as enhancement any modification applied to astronauts that adapts them to the unique space environment and thus gives them a unique characteristic that no humans possess, at least not to the same high degree as a modified astronaut. The problem with this definition, however, is that the health-related enhancements that may be applied to astronauts may differ in the extent of their necessity understood as directly related to survival.

An enhancement aimed at increasing protection against cosmic radiation is incomparably more connected with the increase in chances for survival than the only hypothetically considered moral bioenhancement. The latter, if it were to be applied to space missions, would also be applied with a view to increasing the chances of survival of the population living in the space colony, but in an indirect way, by increasing its capacity for cooperation and minimizing or appropriately channeling antisocial tendencies.

Finally, if we add to this mix enhancements directed at increasing the chances of reproduction in space, or simply making it possible—assuming that it will not necessarily be possible without the enhancements being applied, then we obtain two types of individuals in space, modified and not modified for reproduction. The latter can lead just as good and even better lives despite being in the same place without this specific modification.

Each of these modifications—directly and indirectly related to survival, as well as not related to the survival of the modified individual but related to their ability to reproduce (again, directly if necessary, and indirectly if recommended but not necessary) offers a different kind of effect with respect to the traditional distinction between therapy and enhancement. This distinction shows that adopting a general definition that speaks of applying changes beyond the typical functions of the species explains little here. For each of the four types of changes mentioned will confer functions on the receiving individuals that other individuals do not have. Each of these changes differs from the others in its reference to survival and effects on health. This is especially true of enhancements related to increasing the chances of reproduction in space or simply to make reproduction possible, which are not related at all to individual health and survival.[10]

We can adopt here the evolutionary biological definition of success which is fitness maximization. We can therefore assume that while the aforementioned definition of enhancement may be useful for distinguishing enhancements applied on Earth from specific modifications applied only to participants in space missions, it is useless at the next stage of analysis, when we separate enhancements applied in space from enhancements applied for the Earth environment. For we are then left with at least several different categories of enhancements in space that are more or less related to health and survival, and some of them, those related to reproduction, are not related to them at all.

In summary, the space environment presents some challenges for adequately defining and conceptualizing human enhancement.

Different Motivations for Human Enhancement on Earth and in Space

One difference between human enhancement on Earth and applications for space exploration is the purpose of their application. I assume that the specifics of space missions, their task orientation, and their exclusive nature support the idea that the modifications applied will be health-related, rather than enhancements designed to increase the performance of a function for its own sake, with no direct or at least indirect connection to health.[11]

Even the most trivial and potentially socially undesirable form of human enhancement application, that is, an application aimed at increasing productivity, efficiency, competitiveness,[12] increasing the degree of a given trait that is not at least directly related to health, gets special justification in space as long as we cannot prove that working in a space environment, under conditions of space radiation and altered gravity, can be as productive as on Earth. It is a context of uncertainty that can justify different types of enhancement applied for purposes other than those directly related to health—although I think it is difficult to draw that line clearly for space missions in a way that is probably unique and not comparable to any environment on Earth.

What we would consider unnecessary on Earth (leaving aside the counterarguments based on safety and social equality), such as moral bioenhancement,[13] in space may turn out to be as important to the survival of the mission as protection from cosmic radiation. An inability to evacuate and the awareness of being millions of kilometers from Earth give the space environment a new moral character, which distinguishes it from the military environment on Earth, where evacuation is often possible, and the conflict is constantly taking place on the same planet. A soldier fighting under difficult conditions can at least hope to be rescued in a crisis situation.

Thus, referring to the motivation for acquiring human enhancement for long-term space missions, I am not sure that the space mission environment allows for the introduction of a concept such as human enhancement for trivial purposes that we are familiar with in the discussion of terrestrial applications. Let us assume that medicine enables the application of many so-called trivial enhancements. Let us imagine that it will also be legally permissible to acquire enhancements of a "trivial" (nontherapeutic) enhancements, which, as often presented in bioethical discussions, are likely to further exacerbate social stratification and are associated more often with danger than opportunity. Finally, let us assume that an individual will be able to acquire such enhancement either before the mission or already in the space base, and perhaps receive it during GGE from his parents. If we understand intelligence as "the ability to reason, plan, solve problems (. . .) or learn from experience" (Anomaly 2020, 3), then is not an intelligence enhancement so understood particularly desirable in space, and at least as long as we do not control the space environment the way we have managed to control air travel on Earth in terms of safety.

I am leaving aside here therapeutic enhancements, which may be required for a much longer period than enhancements not geared toward healthcare functions. It is possible that after at least a few years of carrying out long-term space missions with both types of enhancements applied, humanity will decide that it no longer needs to apply enhancement as enhancement (for example, improving intelligence and morality), because she has developed, probably by trial and error, conventional cultural means. Perhaps people from the beginning will apply in the space base some social institutions known from the Earth, some forms that are a resultant of them, which will be considered appropriate to the conditions of the cosmic environment. Perhaps, however, these measures will not prove sufficient and will be assisted by cognitive and/or moral bioenhancement. As long as humans are not evolutionarily adapted to life in a cosmic environment—and the concept of evolutionary mismatch implies that our psyche and morality are not even adapted to life in today's Earth environment— any measure to increase that adaptability should not be morally controversial. And for that reason, it is hard to call it trivial or unnecessary, as in the terrestrial case, where we are usually able to offer working alternatives.[14]

To sum up, the specificity of the space environment gives human enhancement a particularly strong justification, making even seemingly trivial and unnecessary types of human enhancements, which are so perceived on Earth, acquire the rank of necessary or recommended for space missions.

No Problem of Social Stratification in Space

While much of the bioethical argument around human enhancement familiar from current debate could be applied to the space mission environment, certain issues are unique to space missions, as well as certain controversies not applicable in the space mission context. An example of the latter is the division between individual- and group-population based enhancement. I agree with Laura Yenisa Cabrera that we do indeed tend to view human enhancement in terms of individual benefits and the individual's right to self-actualization and self-improvement (Cabrera 2017). However not always, as some emphasize the context of large populations, such as Savulescu and Persson's concept of biomedical moral enhancement. Nonetheless, it must be acknowledged that when we speak of radical human enhancement, we usually mean the enhancement of the individual without any particular attention to the rest of the population. The risk of exacerbating inequality appears here almost automatically.

Interestingly, this seems to be a problem that is unlikely to ever arise in a space colony, or in the initial stages of reasonably short-term missions involving a modified crew. This is an interesting example of a situation in which a serious ethical issue concerning human modification on Earth is unlikely to be applicable in the space environment. This is primarily because, if indeed humanity decides to implement human enhancement at some stage of the mission, it is likely to be mandatory rather than optional. While it may be optional in the sense that someone will participate in the mission at their own risk, there is some reason to doubt that the elite nature of at least the first deep-space missions will allow such a right of choice to be preserved. Their high degree of risk and costliness may make it necessary for political and financial reasons, but also for the success of the mission itself, for each participant to undergo the appropriate modification. Consequently, the specificity of space missions precludes the aforementioned distinction between individual and population perspectives on modification.

We would sooner expect such a stratification in the distant future, where the space base or habitat is inhabited by a relatively large population. But it is also conceivable that life in space settlement will not necessarily enable or require varying degrees of upgrading and enhancement of particular functions and capabilities, beyond the base ones, which may be mandatory for all. It is an open question whether a program of radical human enhancement in space settlement might at some point abandon the framework of what approaches it as therapeutic and preventive purposes and take the form of more trivial enhancement or even extravagant modifications.

However, when I speak of enhancement extravagance in the space environment, I do not mean the same enhancement that passes for extravagance in today's bioethical discussions. I am referring to enhancement that is difficult to imagine, that may be available for space missions, and for which it will be difficult to point out its practical application, its utility from the standpoint of the individual, the group, or the success of the mission, and for which an alternative will be available that is more morally, logistically, and financially justifiable.

The exclusive nature of space missions,[15] as well as the isolation of the members of such a mission from the rest of the population—I am leaving aside the problem of readaptation of astronauts to terrestrial conditions and the risk of irreversibility of enhancements—provide a rationale for applying radical enhancements for the purposes of space exploration. This may be especially true for such a controversial procedure as moral bioenhancement.

In conclusion, one difference between the moral ecology of space and Earth is the expected lack of social stratification in the space environment, which offsets the charges of social inequality made against human enhancement.

Specification in Space Bioethics

Specification is an appropriate methodological framework for bioethical issues in space. Even if we were able to establish some consensus as to the norms that should guide our decision-making regarding, for example, human enhancement for space and reproductive issues,[16] we might consider this consensus to be valid only for the realm of space exploration regardless of what other possible bioethical solutions exist for analogous cases outside the realm of space exploration. But even with respect to space missions alone, it is difficult to speak of a consensus on a single, common position regarding the ethical status of human enhancement.

Different types of space missions have varying degrees of justification, some of which have a rather poor reputation (space tourism), while others are accorded the status of strategic missions for humanity (like missions to save the human species through space colonization), but even these are in practice treated with detachment in the context of terrestrial challenges and no imminent visible catastrophe.

We can envision a situation where in the future we agree to apply human enhancement only to those missions deemed essential to humanity, be it probably not tourist flights, which, as I have suggested, may feature a particularly strong justification for human enhancement. However, if we decide that we do not agree with the application of human enhancement for tourism purposes where it is justified (because we consider the purpose of the endeavor requiring such modification to be trivial), we should abandon tourism missions rather than agree to send unenhanced tourists into space. And it is perfectly possible to imagine a scenario in which we would apply human enhancement at least for the purposes of selected space missions while prohibiting the application of human enhancement on Earth.

As such, specification plays an important role in bioethics in general, not just that of space missions. As James F. Childress rightly points out, even the Christian tradition makes specifications of the scope of the biblical prohibition of killing, from which it distinguishes exceptions (Childress 2009, 70–71). This is an example of a rather radical and inflexible application of the specification rule in morality. The Christian approach to abortion and euthanasia is quite radical.[17] Nevertheless, it can be said that specification is applied at least to the overarching do-not-kill rule, although particular forms of the taking of life, such as abortion and euthanasia, are not exempt from its validity.

I believe that in the case of space missions deviation from certain moral principles and rules may be inevitable, while the challenge remains how to weigh and balance the various conflicting principles and rules in the context of challenges during space missions. I assume that the basic, immutable principles remain well-being and respect for rights-holders. But other equally important principles, such as freedom and autonomy,[18] may in certain situations be suspended in favor of the well-being principle.[19] We should not block the application of human enhancement in the name of defending principles that are by definition unclear and dependent on context, such as the principle of justice and respect for human dignity (in the sense in which it is understood by bioconservatives and serves to reject any human enhancement).

In summary, specification is an indispensable methodological approach in space bioethics for weighing methods and balancing principles and moral rules when analyzing particular biomedical situations during space missions.

The Rule of Informed Consent and Autonomy in Space

The moral rule of informed consent is rooted in both the principle of autonomy (duty-based principle) and the principle of beneficence (consequence-maximizing principle). The rule of informed consent is a key biomedical rule. Interestingly, however, the space environment is one of the rare moral environments (moral ecologies) in which this rule can be easily ignored or simply impossible to implement for objective reasons.[20] And we undoubtedly have some grounds for arguing that, at least in certain contexts and under certain conditions, it should be ignored. Before turning our attention to what status the concept of informed consent might play in relation to future space missions involving human enhancement, it is worth noting what the status of the concept has been in missions to date. Both NASA's Gemini and Apollo missions conducted biomedical research with astronauts participating in those missions. Importantly, participation in these studies was not voluntary, so the criterion of informed consent was not applied here (Sawin 2017).

The space mission environment generates a number of problems that, however acceptable they may be (and in fact they are, for the reason that these problems are acceptable despite violating duty-based principles) from the perspective of consequence-maximizing principles, nevertheless violate duty-based principles. Autonomy and privacy of the medical data of astronauts are violated as long as the data collected from astronaut surveys and observations, which we need to assess the riskiness of space missions, are then used by the scientific community (Robinson 2004). The problem, then, is both the risk of creating conditions of at least indirect pressure to consent and the fact that the astronaut's identity may relatively easily be revealed from a description of their medical condition. But, as Walter M. Robinson points out, the privacy concern is also that the astronauts' sharing of all medical information with the flight surgeon could result in the astronaut's exclusion from participation in future missions. The stakes are very high, so an astronaut may consider the possibility of concealing certain symptoms.[21]

But another risk associated with the issue of informed consent is the right of astronauts to refuse when, in any case, the total sample possible to study is very small, and each subsequent refusal further reduces the effectiveness of the study and hinders the possibility of obtaining knowledge relevant to improving safety in space missions. The problem is that astronauts are the only research group available for space medicine, and unlike studies on Earth samples, it is impossible to find replacement volunteers to participate in these studies (Wolpe 2005).

Robinson points out that NASA's informed consent protocols do not work because they effectively force astronauts to agree to participate in required research during a space mission, because NASA has the right to suspend an astronaut's participation if the astronaut does not agree to participate in research considered essential to the mission (Ball and Evans 2001, 183). Robinson proposes replacing the Common Rule with an occupational model of data collection (Robinson 2004).[22] This problem is addressed by Scott Kelly, who points out that participation in an experiment during a space mission should be viewed in terms of a professional obligation, and that in principle an astronaut should agree to participate in such experiments given their validity, although Kelly emphasizes the voluntary nature of participation and the validity of both the informed consent protocol and the principle of autonomy (An Extraordinary Astronaut 2020). We can add that the high competitiveness and the small number of space missions may create pressure to accept radical moral enhancement in those astronauts who may not otherwise have chosen to undergo enhancement.

We already have some starting point for understanding the ethical value of the rule of informed consent in relation to space missions. The formal application of informed consent does not yet ensure that all ideal conditions are met, which abstractly should be fulfilled in order to speak of authentic informed consent. But is not the actual impossibility of maintaining all these ideal criteria in practice consistent with common sense? Assuming that participation in them is not compulsory, it can be assumed that the participant should accept participation in research that had to be done anyway.

With regard to the peculiarities of space missions, it may be added that their costliness concerning not only the financial outlay itself, but also the time-consuming and labor-intensive nature of crew preparation and training, may compel and justify a little sacrifice that may apply to mandatory participation in research. On the other hand, mere presence in a state of microgravity and increased cosmic ray exposure is just such an experiment, which in a sense accomplishes itself. It is therefore not a separate transition from state A to state B. The space mission itself is a biomedical experiment. In this light—that initial decision to relocate to a dangerous environment, which the future astronaut probably had to make even before choosing her educational path—the need to undergo human enhancement can be seen as a consequence of that decision.

If we assume that bioethics of space missions is similar—although not identical—to military bioethics, we find parallels concerning the limitations in autonomy, which is the source of the concept of informed consent, between astronauts, and in the further future space colonizers and soldiers.[23] The autonomy of space mission personnel is and must be limited. They therefore constitute, like soldiers, a vulnerable community which, because of their subordination to the service and their orientation to the good of the mission, may be forced to submit to certain procedures. Besides, the participants in the mission themselves may feel such pressure (Latheef and Henschke 2020).

An interesting argument for the obligation to accept radical cognitive enhancement is the argument from the advantage in terms of a given trait possessed by the same individual after modification. This argument refers to the superior position of the same individual at some point in time in the future in which, after modification, they will possess decision authority greater than that of the same individual at the time of the modification decision (Latheef and Henschke 2020). The argument allows for the possibility that the individual may wish to object to the modification, but does not have the right to do so as long as the modification guarantees them a higher authority that, if they had possessed at the time of deciding to modify, the decision would undoubtedly have to be positive. The point, then, is that individuals enable themselves to achieve a more perfect state.

Although the argument discussed by Sahar Latheef and Adam Henschke refers to a scenario that is unrealistic today, it can easily be applied to more realistic and directly oriented to the survival of the individual physiological modifications. The authors' example concerns increasing the degree of autonomy in decision-making. Similarly, gaining immunity to certain hazards in space makes the individual considered as a candidate for modification better prepared for space missions and justifies the cognitive situation in which the individual should discern the good of such modification.

The full parallel between the exemplary situations is made difficult by the fact that in the first case the object of modification is to be the autonomy of the individual, which is undoubtedly a high value in itself. In the second case, what is at stake is the modification of a particular physiological parameter, and here the individual could theoretically refuse such a modification by claiming that its value is not improved beyond its better adaptation to a particular situation. Surely the moral value of her right to refuse modification depends in part on whether she must participate in the mission, and what the purpose of the mission itself is.

In contrast, a strong moral obligation to consent can be derived from the concept of moral bioenhancement, which can be treated as our duty, subject to safety and efficacy criteria. It would be difficult to justify the refusal to become morally superior, especially in such specific and demanding environments as the battlefield and space missions. In the case of the concept of abolishing the right to informed consent for the purposes of moral bioenhancement, the analogy between the two seems justified. Such an approach, however logically conclusive it may be, raises not only the risk of paternalism but even authoritarianism. For it could lead to a situation in which some individual superior to the astronaut not only decides what physiological modifications she should adopt, which we might be ready to accept given the challenges of the space environment, but would also decide what is best for her in terms of shaping her morality or cognition in the broadest sense. We cannot rule out a situation in which a committee of medics and scientists "knows best" what behavioral, moral, and cognitive modifications to apply.[24] But equally, at least as long as the missions are militarized or commercial in nature, any range of such modifications may be greater than would be required by objective environmental factors.

In summary, current space missions are familiar with situations where the context of applying informed consent does not meet the expected moral criteria. A problematic proposal seems to be the option of making decisions about applying human enhancements by third parties, which may even exclude the possibility of real informed consent. This is a morally unacceptable scenario.

Space Mission Environment as a Challenge to Bioethical Principles

While the fact of obedience and hierarchy, participation in high-risk missions,[25] and a degree of isolation from the rest of society are common to both space missions and military missions and research expeditions in the Arctic, for example, at least one element distinguishes space missions from the others mentioned above. This feature is the distance from Earth precluding the possibility of immediate return to Earth. One can add to this uniqueness permanent environmental factors such as cosmic radiation and altered gravity that have physiological, psychological, and behavioral effects. But we can imagine scenarios in which, in a given Earth environment, individuals will operate under high radiation, and other harsh environmental conditions (however hard it is to imagine anything corresponding to altered gravity).

In any case, while we can find equivalents of the harsh environmental conditions of space on Earth, as long as we stay on Earth even under the worst conditions, nothing will correspond to the situation of being millions of kilometers outside Earth with all the effects on communications and the inability to return quickly that this causes. I assume that such a return to a safe base and safe evacuation may be possible for military and research missions. Exploration missions beyond Earth orbit preclude the possibility of an emergency return to Earth.

Charles Sawin lists some of the fundamental bioethical challenges that grow out of this difference in location and distance from Earth, thus accounting for the uniqueness of the bioethical environment of space missions. The space environment challenges bioethical principles and rules, not least the autonomy and informed consent. Sawin points out that the status of the rule of confidentiality is expected to change during exploration missions. This may be due to the physician's obligations to the space agency, and indirectly to humanity as a whole, or at least to the state carrying out the mission, if the astronaut's condition does not meet the required criteria. In this respect, however, the space mission environment is not unique, because in the army the confidentiality rule is also limited in the name of the good of the institution.[26]

A similar character, analogous to research expeditions and military missions, is present in another characteristic mentioned by Sawin, namely, the fact that the same individual fulfills different roles. During a space mission, the same astronaut will be at the same time a doctor and a patient, the performer and the subject of experiments. The lack of the possibility of emergency medical evacuation becomes additionally troublesome in the face of the limited medical potential of the spacecraft or space base. Not only limited medical resources, as well as limited personnel, but also limitations in the ability to perform all those medical procedures, including long-term therapy, convalescence, and rehabilitation, that would be available in the astronauts' home countries. Sawin also points to the communication lag unique to the space environment, which can be troublesome in the event of emergencies and telemedicine support.

What is unique to the space environment is the presence of cosmic radiation and microgravity as well as the risk of negative psychological effects caused by isolation compounded by distance from earth. These elements are also pointed out by Sawin when he emphasizes that radiation protection is possible in space only to a limited extent. Hence my proposed concept of human enhancement, treated analogously to that of Sparrow cited earlier, for which GGE may be required by the principles of beneficence and nonmaleficence.

Other bioethical challenge factors cited by Sawin include limitations on diet and exercise opportunities during flight as well as presumably while on the space base—likely significantly limited by altered gravity, among other factors. The latter is a particular challenge, since taking care of one's health and fitness is undoubtedly not only one of the basic rights, but is also recommended. The space environment should not preclude and significantly limit one's ability to take care of one's physical fitness, since such taking care is intrinsically integrated with a good, healthy, and long life, and thus the lack of such exercise available on Earth limits one's chances for survival and well-being—and therefore harms one's most basic interests.

Another group of bioethical problems in space associated with breaches of confidentiality is screening. Sawin gives the example of genetic screening tests. There is little chance of guaranteeing anonymity if only because the population being screened is small, and because the screening will influence personnel policies regarding candidate selection and continued cooperation with the space agency. But analogous challenges will be raised by the application of all those human enhancements that will be voluntary. We can imagine a situation where preference will be given to astronauts with all possible enhancements even when they are not mandatory. Knowledge of an astronaut's medical condition, including the genetic modifications they possess, can be treated as an element of public safety that should be available to those making decisions relevant to the mission. An analogous controversy will arise when access to such knowledge becomes a right or privilege of the organizers and owners of space mining companies.

Sawin concludes his brief report with the following three questions. One concerns the potential mandatory nature of appendectomy to prevent appendicitis in space. Another question concerns selection by blood type to ensure blood group compatibility among the crew for medical purposes. The last question asks about the selection of crew by their gender to ensure the best performance and unity of the group (Sawin 2017). While these proposals would be considered either too self-sacrificing or too discriminatory, the specific context of the space mission environment makes it legitimate to pose such questions even in democratic liberal societies renowned for their antidiscrimination.

We can analyze many scenarios of different types of space missions in different configurations, at different stages of advancement, and with different justifications, and then evaluate the applicability (and validity) of particular bioethical principles. Here, I would like to highlight just one of the principles, the principle of beneficence.

The principle of positive beneficence in its classical formulation implies, among other things, such obligations as the duty to protect the rights of others, to prevent harm to others, and to eliminate conditions that harm others (Beauchamp and Childress 2013, 204). The principle of beneficence requires that the welfare of astronauts be taken care of. One of the tasks for this purpose is to equip the spacecraft with all the necessary medical supplies, as well as to provide adequate medical preparation for the astronauts. But the problem can be the ability to anticipate all possible medical events and the ability to take full equipment due to weight and space limitations in the spacecraft (Wolpe 2005).

If we accept, as Robinson considers, the rule of wartime triage as an exemplary model for space mission ethics, then the criterion for allocating medical resources is the injured person's ability to return to duty.[27] Robinson suggests that application of such a rule would require mutual consent between the astronaut and the space agency. Robinson also raises another important issue, namely the internationalization of the crew and the fact that, as a consequence, crew members may represent a variety of ethical approaches to the triage issue (Robinson 2004).[28] This is an important issue regarding crew selection, which perhaps should also take into account cultural elements such as the religious sympathies of the candidates.

Human enhancement understood from a welfarist perspective obtains a strong justification for space missions. This is particularly evident in light of the principle of beneficence. Human enhancement is a means to protect the rights of others, such as the rights to health and life, but also to prevent harm. Moreover, it becomes practical in light of the challenge cited by Paul Root Wolpe of being able to predict medical accidents and provide all the medical resources that may be needed during a mission. Human enhancement in space becomes a form of prevention.

While this approach may be acceptable for those enhancements that have the direct effect of increasing resistance to cosmic radiation and the effects of altered gravity and are applied only to the individuals involved, it is controversial to apply this type of reasoning to GGE. Since there are already people living in space (in a hypothetical future colony), applying GGE to their future children does not necessarily meet the criteria of the concept of prevention assumed by the principle of beneficence. GGE in a space colony could be treated as a kind of prevention when adults will reproduce soon after arriving in the colony, or at least early enough that space medicine will not be able to confirm whether it is safe to give birth to children in space without GGE. Only proving that GGE is necessary to eliminate some type of harm to future children can justify its application in positive beneficence prevention terms.

But this still does not preclude the application of GGE even in those situations where space medicine does not confirm the need for GGE to avoid harm.

First, harm can be defined broadly, and it is worthwhile to have precisely that broad definition applied in space. Possible harms include the harm associated with psychological trauma that might occur in future children in space, and the harm associated with the loss of opportunities if moral bioenhancement were not applied, assuming that it would only be available through GGE.

Second, even if GGE fails to meet the criteria of the beneficence principle, its application can still be justified by the welfarist concept of human life and gene editing. It is better to have GGE than not to have it in such a demanding environment as space even if space medicine has not proven that GGE is necessary to prevent harm.

In summary, the space mission environment challenges the applicability of certain principles and rules. But at the same time, it imposes special obligations to follow certain principles, like the principle of beneficence.

Space Bioethics versus Military Ethics

What most closely aligns the concept of human enhancement for space missions with military enhancement applications is its preventive nature. If we apply this term to the classical division between therapy and enhancement, we may see some difficulties related to the lack of precise and unambiguous divisions. A solution may be to recognize the existence of a separate category of enhancement, preventive enhancements (Erler and Müller 2021). However, even this distinction is not entirely simple, because within it we must distinguish preventive enhancement in those in whom, in the absence of preventive enhancement, a given condition considered to be unfavorable and exposure to an unfavorable factor will certainly develop, from those in whom it is not known whether exposure to a threatening factor will actually occur. An example of the first group is GGE, when a gene is found that will inevitably lead to the development of a disease if not removed or modified, from, for example, vaccination for a disease for which the risk of incidence is not very high. The latter category is more nuanced, distinguishing vaccination against diseases that can be severe and even fatal from diseases that are usually mild and have a low mortality rate.

An example of preventive enhancement currently applied for military purposes is anthrax vaccination (Erler and Müller 2021). A technically feasible application today could be CRISPR/Cas 9 applied also to enhance resistance to various biological weapons (Greene and Master 2018). Analogous status seems to be given to all types of enhancements that can be considered for future space missions. As for some, one can be sure that they will be useful, as in the case of GGE against a severe genetic disease. Others may be applied with a high standard of precaution, and resemble more a vaccination against a not very dangerous disease. Their preventive nature makes them closer to therapy than enhancement, and in fact justifies considering them as a subset of therapeutic (health-referenced enhancement) interventions.

While the introduction of the category of "preventive enhancement" into the conceptual scheme is arguably helpful in classifying many cases at the intersection of enhancement and therapy, a new and specific conceptual problem arises here. Namely, in relation to prevention, it is necessary to determine to what extent the applied modification is necessary for the preservation of health and even life, to what extent it is optional—and to determine here the degree of risk of occurrence—as well as, finally, its direct or only indirect character. This distinction is important for rationalizing the discussion of human enhancement for space missions.

Also, this distinction should serve the philosophical debate about human enhancement in relation to Earth, when philosophers consider the possibility of extending human life by, say, several hundred years. The rationalization in the discussion of the status of human enhancements for space missions is that only those modifications that are necessary to protect the health and lives of astronauts are seriously discussed. The issue of performance enhancement becomes a certain problem, but even here the consideration of enhancements for space missions is to prevent or compensate for the loss of performance, not to raise it to some super level far beyond standard human capabilities. This rationalistic approach presented in this book is the reason I do not consider speculative forms of enhancements not only because today they seem unrealistic from the point of view of science, but also because, regardless of their technical feasibility, they seem unrelated, or only distantly and loosely, to the chances for survival or just a good life.

The context of space missions shows that therapeutic modifications do not always have the highest risk-benefit ratio. The main rationale for this in many cases true claim of a higher risk-benefit ratio for therapeutic modifications is the belief that more benefit is produced by restoring the ability of a poorly sighted person to see well than by improving the degree of vision of an already well (normal) sighted person (Erler and Müller 2021). To undermine this claim, however, requires a fair amount of variation in the population of those traits that are being modified. This is the element that can significantly differentiate human enhancement in space from military human enhancements.

Suppose we want to improve in healthy and well-trained soldiers, for example, the degree of their accuracy in shooting at a target. Despite the same training, there is a variation even among the best soldiers. This seems true, because otherwise we would have a hard time explaining the fact that snipers with at least apparently unique abilities appear from time to time in history. Thus, in the human population there is a diversity of predispositions to be a sniper, which is not eliminated even by undergoing the same cycle of training.

An analogous situation is not found in the case of human enhancement for space missions. I am referring to preventive enhancements directly related to the protection of astronauts' health and life. The factors generating pressure to create and then apply appropriate preventive enhancements are mainly microgravity and space radiation. Their permanent impact on the human body in space is so great that we do not have the diversity in the population that would make us able to select candidates with greater resistance to these factors. Their greater resilience would, in turn, lead to a situation where they would either not need to be modified at all, or to a lesser degree than other less naturally resistant microgravity and/or space radiation candidates.

Since such diversity in terms of resilience to these factors does not exist, or possibly exists but is insignificant for such long and distant space missions, each candidate must undergo the same level of invasive preventive enhancement. The exception to this is the moral bioenhancement hypothesis, where we know that there is now diversity in the human population regarding differing degrees of resistance to stress and isolation. This diversity makes it possible for moral bioenhancement to be applied to different degrees to different candidates. The controversial nature of moral bioenhancement may justify a special selection procedure only for those candidates who exhibit the highest possible degree of such resilience in the population.

The risk-benefit ratio for the difference between different degrees of predisposition does not play a role for enhancement for space missions, as opposed to military applications, while it can only play a role for moral bioenhancements. However, to establish or exclude an analogy between space and military missions would require an analysis of all modifications that are potentially desirable in soldiers and verification of which ones may occur in a portion of the population as part of natural variation, which ones may be acquired by nonbiomedical methods, and finally which ones may be obtained only through biomedical enhancement.

A parallel to military ethics may be the ethos of the astronaut, associated with particular risks to health and life. If there is a consensus on the ethos of a soldier's profession, who is willing to give their life not only for their colleagues but also for the nation, an analogous situation may exist for deep-space astronauts.[29] In the latter case, however, again justification for the space mission plays a role. If the mission is scientific in nature, shall we say that the astronaut is willing to give their life for the advancement of astrobiology?

If the missions will be commercial in nature, is it reasonable to say that the astronaut gives their life for the profits of their company? Or simply for the profit that their family will at least receive, as in the case of military mercenaries? It is worth bearing in mind the different types of motivation, such as financial motivation, seeking one's own prestige, and acting for the good or glory of the nation or humanity. Until we can draw these lines, the analogy with the soldier's ethos can be difficult.

On the other hand, it may be worth considering the selection of a future generation of deep-space astronauts in terms of having such an ethos. This is relevant to the issue of informed consent and autonomy because it not only makes it easier to implement actions that reduce astronaut autonomy but also removes, at least in part, the risk of opposition in the name of bioethical principles, like autonomy and justice.

Despite the important similarities between the moral ecology of space missions and the moral ecology of the battlefield and functioning within army structures—which we seem to perceive intuitively, there are also some important differences. It seems that a greater degree of restrictions on autonomy and the importance of the rule of informed consent, as well as the correlated greater right to apply radical forms of human enhancement, applies to military ethics than to space bioethics. This is related to the fact that soldiers carry out tasks in which not only can they kill others, but they themselves can also die relatively easily. This specific situation makes modifications that can both serve to protect their own lives and make their decisions about who, where, and when to kill more precise and justifiable desirable, and perhaps even morally obligatory (Latheef and Henschke 2020).

The context that reinforces this argument is the situation of fighting in defense of others, where the good at stake is theoretically the life or independence of an entire nation. This argument can only work against certain soldiers and only in specific situations. However, this does not change the fact that it is easy to imagine situations in which decisions and skills regarding the killing of others require constant improvement.

Space missions are devoid of such a situation. Some analogies can be found but they seem too distant, and a bit stretched. It can be said that controlling the landing of a spacecraft on the approach to landing on Mars after several months of flight in microgravity requires excellent cognitive and physical fitness. Otherwise, the spacecraft is in danger of crashing (assuming that at least some of the strategic maneuvers must be done manually). If a small mistake caused by a grueling flight can cause a catastrophe, and only human enhancement can guarantee a minimum of good fitness, then it becomes required.

Analogous questions can be asked about many other situations in which the mission commander and any of the members must not only be minimally physically fit, but also have some cognitive and moral fitness. If one could find such situations in interplanetary flight and space base as in the military ethics example discussed above, then the modification in question—whether physiological, cognitive, or moral—becomes mandatory. The problem, however, is that it is probably impossible to predict their necessity prior to such a mission on the basis of near 100 percent certainty that characterizes the aforementioned military ethics situations. The fundamental question is if the space mission environment will contain life and death decisions analogous to the military action environment.

These decisions will not involve direct decisions to kill, as in the case of military ethics, but they may indirectly involve decisions affecting crew survival. Such an example could be a situation where an equipment operator on a spacecraft or space base may have limited attention and perceptiveness, and the lives of other members performing a task may depend on her mistakes or slow decisions. In such a situation, there could be some parallels between space and military missions, but provided that we first establish that human enhancement is necessary to maintain an adequate level of cognitive or manual ability. This is a stronger version of the argument. The weaker version may justify such modification even if it is known that radical human enhancement will certainly increase an astronaut's potential and performance, but it is not known whether such an intervention will actually be necessary (whereas we assume that it will be known not to be harmful).

The purpose and context of the possible application of human enhancements is not always analogous for military and space missions. Even within military missions, which we have found to be more likely to justify limitations on the autonomy and scope of applicability of informed consent than space missions, there are differences that affect the ethically acceptable limitations on soldiers' autonomy.

In contrast, it is worth emphasizing that the environment of space missions need not be analogous to that of military operations to obtain a moral justification for human enhancement where military ethics warrants such a justification, as in the aforementioned cases of killing decisions. While some distant analogies can be drawn with respect to situations in which the decision and action of one individual may determine the fate of others in space, space bioethics draws justification from other sources, namely, from prevention under the principle of beneficence, and, where the criterion of prevention is not applicable, from the welfarist concept of human enhancement.

I believe, however, that the aforementioned case of military ethics and the context of killing is an interesting point of reference for space bioethics. Where decisions are made that can theoretically have a direct impact on human health and life—and the space environment makes it so that, in principle, almost any decision can be viewed in this context—it would be acceptable to apply human enhancement whenever its positive effect on improving the physiological, psychological, and behavioral parameters will increase the likelihood of optimal decision-making. This opens the way for considering moral bioenhancement and cognitive enhancement for space missions as reasonable options.

In conclusion, despite the occurrence of some differences between the ethical environment of space missions and of the battlefield, it can be said that military ethics is the closest analog and reference point for space bioethics.

Group versus Individual Interests during Space Missions

The final issue I want to address here is the question of the possible primacy of the good of the group over the good and interest of the individual. It seems that the risk of sacrificing the individual for the variously understood good of the group is high under the conditions of a space mission. I object to the acceptance of utilitarian thinking in space bioethics, which emphasizes only the sum of happiness, which also allows for the suffering of individuals, as well as would allow for the low standard of living of a large sum of individuals in, for example, a space colony.[30] The task of space bioethics is to protect us from such scenarios, which may even unconsciously set in motion such a sequence of events that in the future will lead at some point to a situation in which the sacrifice of an individual(s) will be inevitable for the accomplishment of a space mission. This is one of those situations where I am willing to accept that certain duty-based principles of deontological ethics should be absolute.[31]

In principle, I believe that within space bioethics we should respect all principles equally at the starting point,[32] both duty-based principles and consequence-maximizing principles, but depending on the situation, in accordance with the idea of reflective equilibrium, we should dynamically and flexibly evaluate each individual biomedical situation in space. What argues against duty-based principles is that they often have unclear meanings. We can say that the situation in which even indirect pressure and the creation of conditions in which a future deep-space astronaut would feel pressured to take a decision that would limit or abolish her autonomy or freedom are unacceptable.[33] But on the other hand, the principle of autonomy should not be interpreted as an absolute principle, because in many situations on Earth our autonomy is limited in many ways every day by the well-being principle.

Human enhancement in this context serves the elimination of situations threatening the mentioned duty-based principles. The application of human enhancement protecting against cosmic radiation and the effects of altered gravity reduces the risk of a fatal accident in a situation where there are strong reasons for carrying out a space mission in the conditions of uncertainty and risk. Awareness of the availability of effective human enhancement, both health- and performance-oriented, creates conditions for an autonomous decision to participate in a mission or to abandon it, where the context of risk as a factor influencing the decision will be minimized. Finally, as we will see in the next chapter, I believe that, at least on a purely theoretical level, the concept of moral bioenhancement applied to space missions, especially in the long-term perspective of a space colony, makes it possible to eliminate the scenario criticized by space philosophers (Schwartz 2020), (Green 2021), who warn of the risk of creating a totalitarian society.

Is space bioethics ontologically distinct from bioethical issues related to issues on Earth? It seems not. Instead, it is an extreme version of Earth bioethics of extreme situations like military missions and Arctic expeditions. Space bioethics describes a unique environment that combines various elements of extreme terrestrial environments, but which only occur together in space. It can be said that the main difference between the two is that certain challenges in the space environment will occur permanently, not just temporarily, making the application of certain biomodifications a necessity, and at the very least, the individuals making the biomodification decisions may be under particular pressure due to the impossibility of changing the environment or the unlikelihood that conditions will change.

This fact has implications for our perception of the moral controversy of biomedical procedures. Space bioethics is in some respects analogous to military bioethics. Although the differences between peacetime and wartime medical ethics are so significant that, in principle, one can speak of a qualitatively new ethics (Gross 2006). Like military ethics, space bioethics suggests that we deal with the environment and with phenomena that require a changed approach. Perhaps what is qualitatively different from each other is peacetime medical ethics on the one hand, and battlefield ethics and space bioethics on the other. What distinguishes space bioethics is that it is more turned toward consequentialism, so that it is open to context-dependent assessments of the present value and need for particular duty-based principles, and even ready to reject them.

In summary, space bioethics is dependent on the context of the biomedical problem being analyzed and is based on specification and balancing between principles and rules.

7
Moral Bioenhancement in Long-Term Space Missions

Introduction

The idea of moral bioenhancement is broadly discussed by ethicists and philosophers (Carman 2021), (Gibson 2021), (Rakić 2021a, 2021b), (Rueda 2021). There are good reasons to modify human moral behavioral patterns and possibly moral beliefs. However, while such discussion covers many different issues on Earth, no one takes into account a context of future human space missions.[1] In this chapter, I discuss bioethics of biomedical moral enhancement for space. I assume that we have strong reasons—possibly stronger than in relation to terrestrial matters—to enhance morality of future deep-space astronauts and possibly space settlers.[2] I also show that ethical context of moral bioenhancement for space differs from analogical context on Earth.

The purpose of this chapter is to reflect on the very concept of biomedical morality modification in a specific context not previously considered in the literature, that is, the space mission environment, regardless of what, how, or by what means will be modified to achieve it. I am interested in considering a kind of thought experiment: how moral bioenhancement might work in a space mission setting, whether it is justified or not, and what its application in space might bring to current discussions of moral bioenhancement on Earth. I believe that the possible acceptance of moral bioenhancement should be limited to particular challenging circumstances. Thus, moral bioenhancement differs from the other types of enhancements considered so far, which are viewed as a mission default. But before we move on to space, let us start with Earth.

Introduction to the Concept of Moral Bioenhancement

It is worth starting with a skeptical note. As neuroscience experts point out, "pharmacological optimism" and "neuroenhancement prevalence" hypotheses should be rejected. As they point out, not only is it difficult to develop neuroenhancement for a healthy person that could improve their daily activities. It is even difficult to develop treatments for sick people (Schleim and Quednow 2018). Biomedical moral

The Bioethics of Space Exploration. Konrad Szocik, Oxford University Press. © Oxford University Press 2023.
DOI: 10.1093/oso/9780197628478.003.0007

enhancement is thus a specific form of human enhancement, more controversial and, it seems, not only less likely but also not necessarily required. Perhaps moral bioenhancement will never be applied either on Earth or in space. Perhaps by the time of the first long-duration and deep-space missions other types of moral enhancement— but not bioenhancement—will have been developed that will render the concept of moral bioenhancement unfounded.

In light of the aforementioned neurosciences' skepticism regarding pharmacological optimism, it is hard to disagree with the conclusion of Stephan Schleim and Boris B. Quednow (2018) that instead of considering the idea of biomedical moral enhancement, it is better to take action to improve environmental and social factors so that they positively influence the well-being of healthy people. Because we are talking about moral modification of healthy people here. The issue of how to treat people with certain moral and cognitive deficits is another important issue, but is not discussed here.

Another skeptical point is that it is not really clear what we are supposed to modify—already regardless of whether we can technically do so—to achieve the desired effect. For we can influence behavior in many ways, at least in theory, by influencing and shaping emotions, thoughts, instincts, desires, imagination, and decision-making processes. What is particularly problematic is the possibility of influencing behavior, for that is what is ultimately at stake—the practical consequences of moral modification. But the issue implies a conviction, characteristic of the eugenic politics of the twentieth century, that behavior is heritable. Are there any moral behaviors, good or bad, that have genetic correlates? Perhaps it is the case that behavior is determined by genes that are too complex, as well as simultaneously determined by complex nongenetic processes, to be controlled by genome editing (Brock 2009, 276). And if so, then the idea of biomedical moral enhancement may be challenged to the extent that we recognize that alongside the biological component of a given behavior there is always an accompanying nonbiological component that is beyond the control of genome editing or drugs.

These ambiguities, as well as the controversial nature of the concept of modifying behavior and morality, lead me to consider the concept of moral bioenhancement only as a moral thought experiment for the sake of argument, and not necessarily as a worthwhile alternative to consider, as in the case of the health-related human enhancements discussed earlier. I do not rule out, as I try to show, that we may have very good logical reasons for applying moral bioenhancement in the cosmos, but what is logically conclusive is not always the best solution in the world of real, living people.[3]

Human enhancement of morality is usually considered in a group context. The group context here means the effort to improve human moral decisions and actions so that they serve the group in terms of either public safety, the promotion of altruistic tendencies or a sense of justice, or all of them at once. The group in question can be either a society or simply humanity as a whole. The task of moral bioenhancement is to solve the problems of group action and group decision-making (among other things, the collective action problem). As such, moral bioenhancement differs from other types of human enhancement, which tend to assume effects on the individual and on the population either not at all or indirectly.

Showing the population effects of human enhancement is often treated as strengthening the argument for or against human enhancement. This does not change the fact that human enhancement by biomedical means makes sense at the population level (Anomaly 2020). Nevertheless, I assume that what distinguishes moral bioenhancement, which by the way is essential to its effectiveness, is that it is intrinsically linked to a population context.

It is also pointed out here that humanity has not developed mechanisms within evolution that generate an almost instinctive or automatic concern for the good of the group/population in a manner analogous to concern for related individuals. This line of thinking is used to justify the idea of improving our morality to confront new problems that not only have been caused by human action as a giant population but also threaten the population and require population-wide action to solve. An example of such a problem is climate change and the concept of modifying our morality so that we begin to see the problem in terms of serious, directly threatening problems (Fanciullo 2020).[4]

If, for the sake of argument, we began to follow this line of thinking, we could see an application for moral biomodification understood in this way also on the grounds of space colonization. We do not know what the psyche and morality of future space colonizers will be like in terms of their ability to think, act, and care collectively for the welfare of the environment and the population. We can surmise, however, that since they will be people who will specifically experience the risk of annihilation through the fact of their personal participation in space colonization for planetary survival, perhaps such modification—even if it were medically possible—will simply not be required.

But it may also be that humans will always have a problem with groupthink and action. In this specific context, moral bioenhancement for climate and environmental needs could also be applicable in space. But probably only in a colonization mission, where population numbers will become a problematic factor and we will face the risk of overpopulation and/or environmental devastation. It can be conjectured that in the first stages of space exploration, the number of people, as well as their actual impact on the space environment, may be too small to harm it even if these people did not care about the effects of their actions.

But apart from the very concept of moral bioenhancement understood as it is usually presented in the literature, namely, as pharmacological enhancement, it is worth mentioning the much more radical form that is mind design. Mind design means both the generation of general-purpose artificial intelligence and radical forms of enhancement of the human brain (Schneider 2019, 1–3). In this chapter, I also discuss other forms considered radical, such as genome editing that takes GGE into account.

In sum, the concept of moral bioenhancement is the type of human enhancement that differs from the others in both its level of skepticism about its feasibility and its population-based rather than individual-based context of impact.

Why the Concept of Moral Bioenhancement Is So Controversial

The arguments of philosophers who criticize the concept of moral bioenhancement focus on their concern about the risk of losing or limiting human freedom and autonomy. Harris, for example, points out that a morally modified individual will not be free to choose in making a moral decision. The boundary between knowledge of morality and the decision to act morally will be blurred (Harris 2011).

Sparrow similarly criticizes the concept of moral bioenhancement. According to Sparrow, after moral biomodification, an individual will probably make right and good moral choices, but they will not be the result of her autonomous judgments. What will motivate such an individual's choices and actions will be her biochemical responses, not her autonomous moral decisions and considerations (Sparrow 2014).

The moral status of moral bioenhancement depends on what is being modified. For example, if moral bioenhancement were to reduce the role of emotions in decision-making, if it were to improve the analytical and synthetic capacities of human reason (where moral bioenhancement is intertwined with cognitive enhancement), or if it were to improve our moral dispositions. Regardless of what the goal of moral bioenhancement might be, it is worth noting that in the case of moral enhancements implemented on Earth over thousands of years, it is precisely the improvement of these elements that we seek to improve, to improve our cognition in the context of moral decisions, and to minimize the potentially negative impact of emotions on moral decisions. In essence, conventional moral enhancement, which we have practiced for thousands of years, involves influencing selected emotions and cognitive elements to achieve a desired behavior.

Because the analogous moral enhancement is known to humanity, but differs in the means used, the philosophical critique of the concept of moral bioenhancement should be viewed differently. What accounts for the moral uniqueness of the means employed—for we may assume that the degree and purpose of modification remain unchanged—is arguably that such modification may be irreversible, as well as that it may, even if reversible, leave the individual no choice. Thus, moral bioenhancement may render the individual inflexible, with no possibility of actual influence over the decision-making process, but merely carrying out, perhaps without her having any influence, the biochemical and motor impulses produced.

But even if this were indeed the case, it would seem that the difference between the artificial, biomedically produced biochemical coercion of moral bioenhancement and, considered as a natural effect of a given model of education, the pedagogical coercion produced by the practice of a given religious doctrine or deep-seated family upbringing would not always be discernible.

Imagine that an individual has a very strongly held moral principle that forbids her to lie. And such a person will never lie, whether motivated by religious belief or by the rooting of this injunction in her psyche dating back to her childhood. Thus, she is functionally limited in her ability to make a moral decision by the aforementioned educational context. Similarly, a person who is subjected to moral biomodification can be functionally limited in her freedom of choice, and will always make decisions that are subordinate to a given moral rule. Consequently, assuming that this is how moral bioenhancement will work, we obtain a moral perfectionist who will always make only morally good decisions, that is, decisions that obey a given moral rule. Such a moral perfectionist is thus no different from a moral perfectionist who makes identical decisions but based on other sources of moral education.

From this point of view, criticizing moral bioenhancement would require us to criticize to the same extent the moral perfectionist who never doubts, never hesitates, and always chooses the moral good. Such a critique is not a critique of moral bioenhancement as such, but of moral perfectionism, by whatever means it was achieved. For, in fact, what philosophical criticism of moral bioenhancement would like to preserve is human uncertainty, human doubt, moral deliberation, the clash of different conceptions, visions, and emotions, which clash can lead to different consequences, not always those that are perceived as morally good. It seems that philosophers value more the freedom to think and decide in moral terms and not the socially good consequences, alternatively they value the freedom to think in moral terms and allow decisions that will not necessarily follow from those thoughts but will be socially good.

On the other hand, if our goal is not moral perfectionism, then we are entitled to criticize moral bioenhancement for always leading us to the same morally good conclusion, whereas we would prefer a scenario that leaves us room either for doubt or for making a different decision motivated by different factors. But if this is the case, we encounter the paradox that a morally perfect person who always makes morally good decisions (whatever that means) is not what we want, and is not what the ideal is from a philosophical point of view, where at the same time the ideal is the idea of the good, at least since Platonic times.

I also believe that the philosophical critique of the concept of moral bioenhancement is overstated. First, the case of moral perfectionism and the assumption that moral bioenhancement will lead to moral perfectionism, but achieved by means other than those hitherto pursued in human history, shows that the critique of moral bioenhancement in terms of the way in which moral enhancement is carried out falls short, for an analogous critique should be made of moral perfectionism itself, irrespective of the means used to achieve it.

Second, moral perfectionism aside, there are many situations in past moral upbringing and training that might be viewed as a kind of programming, of instruction given while still in childhood, of imitation of parental and peer behavior by children and adolescents. Knowledge transmission within cultural evolution can occur vertically or horizontally and may involve oblique transmission (Szocik 2019e). In other words, cultural evolution is precisely such moral enhancement, which often leaves no room for free choice.

Third, with macro- and micro-determinism taken into account, we come to the conclusion that we never have such a thing as a state of free choice or simply freedom *in the sense that is presupposed by the criticisms*. Their model of what practical reason is actually like is a flawed model. Biomodified agents would no more operate as simplistic automata, any more than other agents would do so. Changing the ways in which practical reason is skewed (which it always is) would not actually remove the need for deliberation and for representing situations in one way rather than another. Nor would it introduce a new reliance on physical structures. That reliance has always been there. Each of our states is fixed (in a sense "determined") by a number of macro and micro level processes that we cannot eliminate.

To conclude, in such a situation, it is difficult to recognize the bias of the critics of moral bioenhancement, who recognize that only deliberately implemented moral bioenhancement would reduce or exclude our moral freedom, but a number of other deterministic factors are no longer presented by them as such an obstacle.

Moral Progression Can Happen without Biomedical Moral Enhancement

The concept of moral bioenhancement is not criticized solely because of its potentially threatening nature to the specifics of human morality, that is, the aforementioned charges focusing primarily on the loss of autonomy, freedom, and the establishment of biochemical determinism. Another critical perspective is to question the need for moral bioenhancement due to the belief in the sufficiency of traditional, nonbiomedical means of improving human morality. This is an argument formulated by Powell and Buchanan, among others. These authors emphasize that since the Enlightenment, a process of moral progression began that accelerated after World War II and consisted of an ongoing, essentially slow-motion transition from an exclusivist to an inclusivist morality (Powell and Buchanan 2016).

Institutions play this role, along with their ability to punish morally unacceptable behavior. Influencing individuals by externally forcing them to take desirable actions, and discouraging them from taking undesirable actions, can, over time, lead to the shaping of their motivations, that is, to the actual improvement of their morality, not only through forcing actions—which may, however, be an inevitable starting point, but ultimately through the shaping ("improvement") of their motivations (Powell and Buchanan 2016, 255–257). This is an issue that touches on a huge area of research, namely cultural evolution. I do not take up this thread here because of its uselessness, or possibly insignificance, for the topic I am discussing of future space exploration.[5]

Despite the criticism of the concept of moral bioenhancement in the current literature regarding its potential earthly applications, despite its controversial nature, as well as the facts that there are phenomena of cultural evolution with social institutions at the forefront that often play a good role in moral progression, I believe that it is possible to imagine scenarios on Earth where the application of moral bioenhancement may be justified. I agree here with Anomaly, who while recognizing the value of institutions, insists that they are only as good as the people who work in them (Anomaly 2020, 19–20). If moral bioenhancement is in principle plausible under some terrestrial conditions, then surely it will be plausible under some conditions in space.[6]

Moral Bioenhancement in Space—For All or For None

Let us assume, for the sake of argument, that everyone on the mission will be morally modified. If this were optional, or if such modification were only applied to participants in selected missions—but to the same locations—we would run into a serious legal problem. The problem can also be expressed by the following question: how to treat, from the point of view of the law, people who have undergone moral biomodification in comparison with those who have not undergone such modification? Should they be judged differently in the case of their participation in a criminal act (Shook and Giordano 2017)?

This problem disappears if we assume that biomedical moral enhancement applied to space exploration will be mandatory for all. If so, this accounts for the ethical uniqueness of the space environment caused by the scope of application of a given form of enhancement. We cannot rule out a situation in which biomedical moral enhancement remains optional for a reasonably large population of a future space colony. While physiological enhancements may be applied obligatorily for everyone, biomedical moral enhancement may remain an option, at least for large populations. However, it may also be that, to avoid the risk of having in one population both people who are morally modified and those who are not, mission organizers may decide to standardize on this issue. Consequently, moral bioenhancement may be mandatory for everyone, or prohibited.

Why Even Consider Such an Idea in the Context of Space Missions?

Space Mission Environment as the Best Place for Testing Moral Bioenhancement

It is worth leaving aside for the moment the controversial nature of moral bioenhancement itself, as well as its validity. Instead, it is worth noting the attractiveness of the space mission environment for applying and testing such a procedure before it is applied—if ever—on Earth.

The space mission environment will have two characteristics that occur naturally in space but would require unnatural conditions on Earth. The first feature is the closed nature of the community. The second feature is its complete isolation from the rest of the morally unmodified population. As an additional condition occurring in space, although not constituting a uniqueness of the space environment, one can mention the intense selection pressure for moral improvement, affecting all individuals permanently and equally.[7] These are important and unique features of the space environment, guaranteeing no contact between modified and unmodified individuals—assuming that every member of a space mission would be subjected to moral bioenhancement.

Also, applying moral bioenhancement to such a relatively small, functionally very similar group is simpler and clearer in a technical and tasking sense. Unlike the small community of space missions, the idea of moral bioenhancement discussed for application on Earth already inherently faces serious difficulties and is essentially cumbersome by definition. It is problematic in a technical sense—namely, how can we apply moral bioenhancement to billions of people? Even if we decided that all of humanity on the planet should be modified, we know that such a process would take place gradually, probably much more slowly than the Covid-19 vaccination program. As an aside, if such a relatively simple, safe, and medically justifiable procedure as accepting the Covid-19 vaccine triggers protests around the world, we can only imagine how strong the resistance and suspicion will be to any mass moral bioenhancement program—unless it is implemented without our knowledge.

Such a gradual process of moral bioenhancement could have undesirable side effects. During the application of moral bioenhancement two groups of people would coexist, those who have already received moral bioenhancement and those who are waiting for it. Depending on what would be the effects of the application of moral bioenhancement, and exactly what kind of morally relevant functionality such a modification would affect, we can expect that among the still large group of unmodified people, there would be individuals who would deliberately, or purely as a result of their character, make decisions that would be detrimental to those already modified, who because of the applied modification would be, for example, gentler and even completely free of aggression. There would also be the problem of the law treating two types of populations equally.

But applying moral bioenhancement on a global scale to the Earth poses not only logistical challenges, but ethical and political ones as well. Many of these objections are debated in the current biomedical literature, just to mention such issues as whether it should be mandatory or not, whether overt or covert, and finally how to bridge the inevitable differences between modified and unmodified.

Finally, what about those who, from a public good and safety perspective, might most need such enhancement but are not necessarily interested in applying it? This is another specific characteristic of moral bioenhancement, which on the one hand will be applied to so-called healthy or normal people, but at the same time to people who need it, so it is not an enhancement for so-called trivial needs, however at the same time it is not a therapeutic need. These and other objections make the idea of moral bioenhancement on Earth a vicious circle—even if we were to agree on the purpose, degree, and justification, we will long retain the coexistence of two populations, one not yet modified and one already modified, with all the socially negative consequences. The space mission environment bears out these objections, and as such offers unique opportunities to test and apply this already inherently controversial idea.

As with the risk of unfairness and the threat to social justice that might arise if morally modified humans returned to Earth, I believe that the population of morally modified astronauts will be small enough (and one might presume that the first group of modified astronauts will be subjected to long-term research and observation before moral bioenhancement is to be applied on a wider scale in space) that the mission organizers will provide them with protection from the morally unmodified rest of the population, or vice versa (depending on which one fears more in the case of a mix of morally modified and morally unmodified humans).

Moral Bioenhancement in Space Can Solve the Collective Action Problem and the Problem of Human Enhancement as a Positional Good

One of the biggest barriers to applying moral bioenhancement at the population level—if we are concerned with solving population problems—is the logistical problem known as the collective action problem. This problem involves synchronization, logistics, but also mutual trust and the willingness of each individual to sacrifice a portion of their well-being (Anomaly 2020, 36). Especially the problem of trusting others, complete strangers, that they will not use our weaknesses (because after applying moral bioenhancement we should not be so aggressive and belligerent) against us.

As Peter Singer rightly points out, perhaps everyone would like to live in a world made up exclusively of people who have less aggression and more altruism, but until the culture changes to the point where these traits facilitate rather than hinder success, parents will probably prefer to equip their children with at least a little aggression and selfishness to enable them to succeed in the world (Singer 2009, 288; see also Anomaly 2020, 26).

Suppose, however, that we live in space settlement and agree that we need biomedical moral enhancement for various reasons. At least three features of the space colony suggest that by applying moral bioenhancement at the population level, we will avoid the aforementioned collective action problem. First, it will be an isolated population where migration will be impossible. Second, it will be a population of relatively small size. Third, it will be a population of people specific in terms of psychological history, shared experiences, and motivations, but also training and selection. As a consequence of these three characteristics, it will be a population that will be, on the one hand, easier to control as well as to monitor. But perhaps because of selection, training, and probably some degree of militarization and discipline, obtaining informed consent as to the application of moral bioenhancement will be easier than on Earth.

The consequence of successful moral bioenhancement at the population level will be the resolution of human enhancement as a positional good to be treated as an intrinsic good. I mean a situation in which the morally modified population will not be interested in treating human enhancement as a tool for gaining advantage over other individuals in the population. We may believe that at some stage in the development of the space colony, when humanity has mastered the environmental conditions sufficiently to produce the social conditions for the emergence of the hierarchical society and social divisions familiar from Earth, moral bioenhancement may be applied precisely for this purpose, to prevent such hierarchization if it is found to be a natural tendency of humans and can be eliminated only by biomodification.

In summary, the environmental limitations of the space environment make it an ideal place to test moral bioenhancement, avoiding serious challenges that are perhaps irremediable when applying moral bioenhancement on Earth.

Correlation between Space Mission Types, Their Justification, and the Need for and Justification of Moral Bioenhancement

There is no doubt that, as in the case of the relationship discussed in chapter 5 between the types of space missions and the justification specific to each and the moral status of various biomedical procedures graded according to the type of mission, a similar relationship will occur with respect to the concept of moral bioenhancement. In the following, I consider this correlation in relation to a two-pronged categorization of different types of space missions. One categorization concerns the division of space missions by goals and motivations, that is, it includes the scientific, commercial, and colonization missions considered in chapter 5.

The second categorization is based on the size of the population of a given mission, regardless of its purpose and nature. The two basic types distinguished here are missions consisting of a small number of members and missions involving a much larger number of people. The latter need not apply only to colonization missions, but also to the two earlier types. The development of means of interplanetary transportation may make it possible to carry out missions for larger numbers of people who will be interested in either scientific or commercial space exploration.

I make an important distinction between moral bioenhancement and other types of enhancement, which I treat as more or less therapeutic or preventive as it relates to health. While, as I showed in previous chapters, I see no ethical objection in principle to the application of even genome editing as long as it remains health- and life-referenced, I allow moral bioenhancement to be accepted only in exceptional situations, only in a select range of space missions.

Moral Bioenhancement in Scientific Missions

While no one has landed on Mars yet, we can venture to say that this will be the type of mission that most closely resembles the longest missions to the ISS to date in terms of behavioral challenges. Uncertainty about the sufficiency of conventional preparation and selection for, for example, a scientific mission to Mars lasting, including a round trip flight, about 3 years is one challenge. Another challenge is the justification for a science mission that would require or justify at least serious consideration of applying moral bioenhancement for the astronauts to participate. The answer depends on the degree of importance we give to the various scientific objectives (cf. chapter 5).

I do not believe, however, that scientific missions justify the application of a procedure as risky as moral bioenhancement, however important it is how it would be applied (whether pharmacological or genetic enhancement). I believe that any scientific mission—except perhaps one in which we can convincingly prove a direct effect on the survival of our species—provides too trivial a reason for moral bioenhancement. This is primarily because moral bioenhancement is not directly related to health and life, which does not change the fact that one participant in a space mission behaving aggressively and uncooperatively can destroy the entire mission.

Still, such an application should depend on the end, and the end in this case does not justify the means. Such an application would be more like an experiment on laboratory animals. To test moral bioenhancement in the field, regardless of the sheer controversy of possible in vivo experiments on humans on Earth, we need to have a kind of "zero astronaut group," which will be the first group of morally modified humans sent into space. We will not know the possible side effects, because even the mentioned possible experiments on humans on Earth do not simulate cosmic radiation, altered gravity, as well as the stress and anxiety of participating in an interplanetary mission and being aware of the distance from Earth. These are conditions impossible to simulate in laboratory conditions on Earth. And the possible scientific benefits are too low a stake to expose astronauts to such risks. To test on the first humans the moral bioenhancement that we would like to apply in space, we would have to not only isolate the study group for a period of 3 years but also simulate the physical conditions mentioned above that are impossible to reproduce on Earth.

Moral Bioenhancement in Commercial Missions

Identical objections apply to commercial missions, whose rationale is much lower than scientific missions. The only exception might be the kind of space mining, where obtaining raw materials in space would be strategic for the survival of humanity on Earth, which seems unlikely (cf. chapter 5). But this is an unrealistic situation not only because it is doubtful that the exploitation and transportation of raw materials from space would be profitable to the extent of saving the survival of humanity on

Earth. It is also bioethically unrealistic for the reason that the first space mining operations will certainly not be considered to save humanity, while they will make it possible to verify the behavioral health of the workers involved in such missions, and the effectiveness of conventional moral enhancements.

By the time humanity reaches that dubious point in the future where space mining would serve its survival, that point will have been preceded by a number of commercial missions that will have already provided data on behavioral health in space. Only then, with the demonstrated ineffectiveness or limited effectiveness of conventional means, could the application of moral bioenhancement for space mining to save humanity be justified.

Moral Bioenhancement in Colonization Missions

This is the type of mission that has the greatest moral justification, and therefore can also potentially justify even controversial means of achieving that goal. For the sake of argument, let us assume the probably unrealistic condition that moral bioenhancement would be first applied only to colonization missions. This is unrealistic insofar as this type of mission is bound to be preceded by the execution of many scientific and commercial missions, where the likelihood of moral bioenhancement being applied at some stage of the mission will be quite high. The following points are worth discussing under these assumptions.

First, because of the risks of applying moral bioenhancement for the first time in vivo to astronauts participating in any space mission, there are good reasons for not applying moral bioenhancement to the first group of people sent. The primary procedure should be the use of conventional means and observation in the field. Only the detection of significant behavioral abnormalities not correctable by conventional means could justify the application of moral bioenhancement.

Someone might rightly object that, despite the controversial nature of applying modifications that affect human behavior without first being fully vetted in the field, such a large project as a colonization mission is too high stakes to risk poor behavioral health and the possible failure of such a mission. True, this is a strong counterargument. But its actual value is weakened by the availability of large amounts of data on the behavioral health of perhaps thousands of participants in long-term scientific and commercial space missions. Such data would certainly make it possible to predict the possible behavioral dynamics of a colonization mission, and would make it possible to refrain from applying moral bioenhancement at least in its initial phase.

Second, imagine a scenario in which too many overly dangerous behavioral anomalies are found to exist at some stage of a space colony's existence, perhaps even in the early stages, that cannot be corrected by conventional moral enhancements. It is worth remembering that a specific limitation of a space colony may be the lack of resources, including humans, who can perform control and discipline functions. But even if it were possible to delegate some of the population to perform these functions, there would still be a risk of abuse of power. I believe that the scenario described justifies the application of moral bioenhancement.

Third, let us complicate the scenario described above a bit and consider whether, after finding the above anomalies, humanity should not send each successive member of the colonization mission with obligatory moral enhancement applied. I believe that it should. However, I also believe that at the same time the humanity on Earth should work on the invention of effective means of conventional moral enhancement and carefully observe and analyze the behavioral health data of the population modified in the space colony.

However, this scenario contains one danger that we wanted to avoid at least in space, that of allowing a modified population to coexist with an unmodified population. I believe that such a situation is more dangerous in space than on Earth because of the isolation, the distance from Earth, and the dependence on a life support system. A totalitarian and even terrorist scenario depends on the effect of moral biomodification in space, whether the astronauts subjected to it lose any capacity for self-defense, whether their trust in the honesty and good intentions of others is boundless. No moral biomodification should interfere so deeply. But where such a risk exists, one might consider a policy of mandatorily applying moral bioenhancement to all participants in colonization missions, starting with "group zero," to avoid moral diversification of the population in space.

Here we encounter a vicious circle in being able to predict and justify such a situation due to the fact that it will only be possible to verify this scenario after sending unmodified humans on a colonization mission. I see three solutions to this problem. The first solution is to send the first group of morally unmodified humans and have them sent back to Earth by the modified humans. The second scenario is the exact opposite, which is to send the modified back by the unmodified in the event that it turns out that moral bioenhancement is not only not needed, but has side effects that only became apparent to the first group in space. Third, we can believe that the available behavioral health data from earlier noncolonization missions will allow the organizers of the colonization mission to draw the appropriate conclusions and determine whether the very first group should be sent out with moral bioenhancement. In any case, I believe that we should avoid mixing morally modified and nonmorally modified people in space.

Fourth, as a final scenario, consider a situation in which we recognize that a population living in a space colony should apply moral bioenhancement by GGE. I believe that this is morally justified on the grounds that the population will have already lived long enough in space to recognize the necessity of moral bioenhancement by GGE. Suppose that for some reason classical moral enhancement will be inadequate, and that secondment of a portion of the population for the purpose of raising and educating children will not be possible. Instead, I see two problems here, definitely practical, logistical, and technical, but not moral.

The first problem concerns the safety condition. This is because this is the situation where for the first time in vivo moral bioenhancement will be applied through the

GGE pathway and we will not be able to predict all the negative side effects. The solution to this problem is a scenario in which humanity, before reaching that point in the future where it is technically and morally capable of applying moral bioenhancement in space, perhaps will first apply moral bioenhancement by GGE on Earth. In such a situation, there would be the possibility of avoiding side effects in the space colony. The problem, however, is that the living conditions on Earth may not justify the application of moral bioenhancement by GGE, and only the space colony could be considered a special moral environment.

The second problem concerns the mixing of populations morally modified by GGE with others not modified by GGE. To avoid this, it would be necessary to modify adults already living in the space colony through SGE and other invasive methods, which is likely to be done in this scenario in which the population in space decides to morally biomodify by GGE. This will be such a critical point that humanity will probably apply all available means of intervention to the adults living in the space colony. In this scenario, we can avoid the risk of mixing a modified population with an unmodified one solely through mandatory somatic modification of adults and GGE of future children.

Moral Bioenhancement versus Population Size in Space

Making a distinction between numbers with small and large numbers of participants, as opposed to human enhancement per se, may be critical to the moral status of moral bioenhancement. I am not arguing that the number of mission members does not play a role for human enhancement. Also, at least to some extent, it can affect the moral justification of applied biomedical procedures. It is enough to emphasize the specificity of missions of a potentially mass character, namely, colonization missions, which can be considered as missions that should be accessible to everyone. Therefore, missions that are accessible for everybody can be considered analogically to the air transport present on Earth, which does not require any passenger, even the type considered as disabled, to undergo special enhancement enabling them to travel by plane.

In many cases, however, mission size generally does not play a significant role in the moral evaluation of human enhancement. The case is different for moral bioenhancement. Moral modification is not directed directly at survival and health issues. It is guided by the context of survival to the extent that good behavior can guarantee the survival of mission members. Thus, influencing the rationale for the need to apply moral bioenhancement to space missions are the intragroup dynamics and behavioral and moral specificities of humans in isolated groups, in extreme expeditions.

Despite the fact that moral bioenhancement may be essential to the success of a mission consisting of only a few individuals, I argue that the mass colonization of space gives moral bioenhancement—under the conditions discussed in the examples

above—a special justification. Such a justification is the assumption that space colonization will be motivated by the desire for the survival of the human species *at all costs*.
At all costs is an important category here for three reasons.

First, the very idea of saving the species through space colonization is an extreme,
decidedly extraordinary measure. If we undertake this task in the future, it will mean
that we must indeed care deeply about the survival of our species. To this end, we are
taking all necessary measures, including applying moral bioenhancement by means
of genome editing, including GGE, because the specific conditions of life in space
make it more prone to catastrophe as a result of misbehaving individuals. It would
be a paradoxical display of irrationality if humanity, after so great an effort as to
evacuate at least part of the population to a space base, neglected then, for example,
in the name of some perhaps misunderstood deontic principle, to apply moral bioenhancement to keep that remnant of survivors in proper moral and behavioral
condition.

Second, a special justification for moral bioenhancement is gained from the fact
that living conditions in space can be unbearable, which can give rise to various behavioral anomalies dangerous to the survival of the population. To avoid this, and
consequently to increase our chances of survival, we should apply moral bioenhancement on a massive scale.

Third, as long as the risks of exploitation and overexploitation accompany the
space colony, the space population should adopt moral bioenhancement for those
purposes for which it is proposed for climate problems on Earth—to prevent ecolog-
ical disaster in space. The above three arguments are special justifications in favor of
applying moral bioenhancement for mass space colonization, or simply in space settlement understood as space refuge.

In sum, the moral justification for applying moral bioenhancement is limited
only to certain specific circumstances considered in relation to the concept of space
colonization.

Moral Bioenhancement for Space Missions Morally Acceptable but Technically and Logistically Problematic

The concept of moral bioenhancement is itself more controversial both morally and
conceptually for the following two reasons. These reasons are equally applicable to
both Earth and space exploration applications.

First, moral modification has something very intimate and personal about it; it
seems to touch what might be called the soul, the psyche, the essence of human being
and humanity, the most intimate sphere that constitutes human identity. For some
critics this is reason enough to oppose such changes. But even proponents of moral
modification should keep in mind the risk of not entirely predictable changes in the

behavior, character, mood, and perhaps to some extent the personality of the modified individual.

Second, the problem with this concept also has a purely technical and operational dimension. Namely, it is not entirely clear what to modify and to what extent. In human life, every individual needs both aggressive and deeply empathetic
behaviors, moments of shyness as well as courage. We do not know today how,
in practice, the profoundly morally interfering action of moral bioenhancement
would consist, beyond the periodic effects of the drugs currently in use. Would it
mean a complete shutdown of a given emotion? Would not an individual be able to
restore it at a given moment, to turn off the action of moral bioenhancement? While
being strong, immune to disease, and resistant to factors of the cosmic environment
does not interfere with everyday life and various situations, being merely "good,"
empathetic, and courageous can cause easily imaginable problems depending on
the place, social relations, situation, task at hand, and the age of the individual (assuming that a given modification, for example, genetic, would already be irreversible and permanent).[8] Nick Bostrom and Rebecca Roache provide a good example
of the reasons why it is difficult, if at all possible, to determine the objective state of
moral and psychological well-being, as opposed to the relatively objectively ascer-
tainable characteristics and parameters of physical human enhancement (Bostrom
and Roache 2008).

In particular, I find this last objection to moral bioenhancement to be the most
serious one, one that is based on a utilitarian yet consequentialist calculation. The
inability to eliminate these doubts, however, should not be an obstacle to the application of moral bioenhancement to space missions. On the contrary. The very idea
of such modification for space missions is desirable from the perspective of mission
success. The precautionary principle has limited application here to counter such
modification. The peculiarities of the space environment make it an excellent place to
apply moral bioenhancement, where the risk of socially undesirable side effects will
be minimal. In this respect, the space mission environment resembles the world of
sports in the context of doping applications. Both environments are hermetic enough
that the application of human enhancements in either environment will not result in
direct effects on other members of the community.

It is worth noting another controversy surrounding the idea of moral bioenhancement, which is not so much an indictment of it (although for bioconservatives it undoubtedly may provide ample grounds for additional rejection of the concept) as
it illustrates the aforementioned high level of controversy surrounding biomedical
modification of morality, much higher than the idea of human enhancement itself.
I am referring here to the concept of cognitive enhancement. I do not consider the
concept of cognitive enhancement for two reasons. First, this level of cognitive enhancement that is currently being pursued is not controversial when it comes to the
application of various attention or concentration enhancing agents. Second, radical
forms of cognitive enhancement that would involve genetic modification—aside

from my caveat about the difficulty of finding genetic correlates of certain cognitive and explicitly intelligence-related functions—are not considered in this book as strategic for survival in the space environment. I do not interpret the space environment as particularly cognitively demanding to the extent that astronauts require radical cognitive enhancement.

Nevertheless, the concept of moral bioenhancement does to some extent evoke the concept of cognitive enhancement, thus increasing the level of controversy surrounding the idea of modifying morality. Perhaps moral bioenhancement cannot be applied without the parallel application of cognitive enhancements.[9] Perhaps it is not enough simply to modify our emotional responses, but such modifications must be accompanied by modifications in cognition and intelligence, for example, in the direction of greater rationality. It is difficult even to foresee here the appropriate direction for such possible cognitive modifications to be pursued for the purposes of parallel moral biomodification. Instead, one can easily imagine that there are many combinations here, and that any direction of modification of both morality and cognitive functions could lead to negative consequences. My point, instead, is that making a person, for example, less aggressive (moral bioenhancement) may require improvements in their analytical, predictive, and contextual understanding abilities (cognitive enhancement) that make her level of aggression smoothly modulated depending on the individual and the context.

I therefore take the position that cognitive enhancement alone is not sufficient to improve morality by biomedical means. Proponents of cognitive enhancement alone might point out that a mere understanding of what suffering is, that all people are equal, not to do to others what we ourselves do not want to receive from others, should suffice for morally correct functioning. However, it seems that behavior, in addition to cognition, also depends to no small extent on emotions. If only for this reason, in order to give such a cognitively and intellectually modified individual the motivation to behave well. Because moral bioenhancement is primarily about creating or increasing in us the motivation to act well. And mere knowledge of what is good, as we know, is not an effective motivating factor.

In conclusion, I believe that the objections to moral bioenhancement for space missions, even by means of GGE, are merely technical and logistical in nature, concerning the problem of confirming the safety of the procedure in the sense of absence of side effects, as well as avoiding the risk of mixing modified and unmodified populations. As I have indicated, these are serious problems, but they are not moral problems (only applying moral bioenhancement without solving these logistical and technical problems will create moral problems), and it seems that they will be avoidable at the planning stage of a colonization mission.

Nevertheless, due to the controversial nature of the very idea of modifying human behavior, I believe that the default procedure for behavioral health during all long-term space missions should be the use of appropriate social and cultural institutions (in addition to appropriate selection and training). Only then should the concept of biomedical moral enhancement be considered to support unconventional measures.

Otherwise, we may overlook an important element and end up, perhaps unintentionally, creating a habitat in space that is barely worth living.[10]

In summary, therefore, the logistical challenges associated with the application of moral bioenhancement, as well as the lack of certainty as to whether biomodification of human morality may actually be necessary for long-term space missions, suggest extreme caution in considering such an option.

8

Space Bioethics, Population Ethics, and Space Colonization

Introduction

In this chapter, I present topics of future people, population ethics, and the problem of intergenerational justice in relation to space mission. I will highlight how the context of space missions, especially long-term space exploration that ultimately presupposes the concept of space colonization, may influence the understanding of these concerns. I devote considerable attention to the philosophy of antinatalism.

Is It Worth Saving Humanity at All Costs through Space Colonization?

I assume that the life of the human species has value in itself. But from this intrinsic value does not follow a duty to protect the life of the human species at all costs and by extraordinary means. At most from this value we can derive an obligation to protect the survival of the human species only by normal means. An analogy is warranted here with the protection of the individual life of the patient, which can be achieved by ordinary but not necessarily extraordinary means. Ordinary measures aimed at protecting the survival of the species include combating the effects of climate change. But colonization of space or a massive human enhancement program aimed at making people immune to some harmful agent is already an extraordinary form. The question is not whether we should strive to save humanity at all, but whether we should think of saving humanity by way of space colonization, being able to anticipate certain limitations and inconveniences that life in a space colony implies. The alternative remains to look for the possibility of salvation only on Earth or, if this is not possible, to accept the annihilation of our species.

There are at least three issues here about our relationship to future humans relevant to space bioethics. The first issue is whether we have an obligation to ensure the survival of future humans by any means, including space colonization. The next issue is whether we are morally justified in shaping the lives of future humans without their consent and knowledge under nonterrestrial conditions when humanity decides to colonize space motivated by protecting the survival of the species. Finally, the third issue is our right to apply GGE to space exploration.

The Bioethics of Space Exploration. Konrad Szocik, Oxford University Press. © Oxford University Press 2023.
DOI: 10.1093/oso/9780197628478.003.0008

There are many ethical and philosophical objections to the idea of saving the human species through space colonization. For the purposes of further discussion, however, I assume that it is worthwhile to consider the concept of space refuge understood either as a way to save the existence of the human species (and other species, by the way), or at least as a way to increase or maintain the good quality of human life.

Population Ethics on Earth and in a Space Colony: The Ethics of the Quality of Life

Assume that the tipping point on Earth has already been passed in terms of the consequentialist link between the total sum of happiness and population size. Consequentialism must, therefore, wish that this growth be slowed down, at least so long as new lives do not increase the total happiness number—on the assumption that the happiness level of each new life is so low and lower than the previous generation that the happiness gain will spoil the total utilitarian outcome for humanity as a whole. But this unfavorable balance can be changed by humanity's expansion into space if each new life born in space has a sufficiently high level of happiness, high enough so that the sum of the happiness of all new lives in space increases the total sum of happiness taking into account the entire Earth population (with a not very high sum of happiness) and the new population in space (with a significantly higher sum of happiness due to access to new resources, unlimited space at least temporarily, and—at least temporarily—no overpopulation). These conditions will probably change as overpopulation increases, but by the time the population in space reaches a critical moment analogous to that already exceeded on Earth, it can grow in an "unlimited" way until a new location in space is discovered.

We can assume that the dynamics of human population growth in space, at least after a certain period of time when such a human colony stabilizes, will match the historical dynamics on Earth whenever humanity begins to reproduce. If, however, due to the harmful space environment, humans cannot reproduce in space (Szocik et al. 2018), the problem of overpopulation disappears, but then the basis for treating space colonization as a way to save our species also disappears. Such a situation could support antinatalism in a perverse way if it turned out that we want to save the currently living generation by sending it to a safe place (a space colony) where it cannot reproduce.

We can therefore assume that the dynamics of human population growth in space, at least after some time, when such a human colony has stabilized, will match the historical dynamics on Earth. Improving the conditions for continued existence would therefore require interference with reproductive autonomy and procreative liberty. Such interference would be justified by the concept of procreative beneficence, but understood in the light of population ethics. In this context, the concept of procreative beneficence does not simply mean the right of parents to design the best characteristics of the child even by biomedical means. It means creating conditions for the future child to live in such a way that the ethics of quality of life recognizes as minimally optimal. Moreover, in the context of population ethics, the relevant party here becomes the human population as a whole, which should wish to have such new members who will share with it the concern for the population as a whole.

To avoid overpopulation, reproductive rights could be at least partially restricted.[1] As Greg Bognar suggests, such a restriction need not lead to a reduction of procreative liberty, but, in line with the idea of liberal paternalism, it would consist in the creation of choices that, in consequence, create a real situation of greater autonomy (Bognar 2019). A kind of compromise between preserving reproductive choice and antinatalism is to place some restrictions on reproductive freedom, but ones that would agree with rule consequentialism, that is, that could be accepted as an ideal set of universally applicable rules. Such a rule might be the "zero reproductive choice rule" considered by Tim Mulgan. According to this rule, an individual can reproduce whenever she wants, but on the condition that the new child will have a life worth living (above the zero level). Such a rule introduces a restriction acceptable to all but those who cannot create a child whose life will be worth living. But, as Mulgan rightly points out, it is difficult to regard such reproduction as a worthwhile life scenario, and it is also difficult to regard such a restriction as a limitation on autonomy when the result of this apparently autonomous action is to bring into existence a human being with a poor quality of life (Mulgan 2006, 167).

Along these lines, if we already agree that we should strive to save humanity by way of space colonization, we should adopt as a working principle the aforementioned rule of the zero level or lexical level reproductive choice, in order to prevent a decrease in the quality of life in space caused by overpopulation, in a manner analogous to that currently experienced on an overpopulated Earth. We do not know whether the space colony would consist only of the most physically and medically prepared individuals, and whether it would also include, for example, people who are sick and in generally weaker condition. The latter could be denied the right to reproduce in such a colony, according to zero level or lexical level rules, or obtain such rights only after undergoing appropriate biomodifications.

In turn, another problem independent of overpopulation, which, however, the situation of overpopulation could obviously only make worse, is the difficulty of living in space, with constant high exposure to cosmic radiation, in difficult conditions of altered gravity, constant stress, enclosed space. It can be said that the quality of such a life will be very low, perhaps a Parfit's "barely worth living" life, where people will simply persist and higher pleasures will simply be impossible to realize.

One of the greatest challenges identified by population ethics is to avoid Parfit's repugnant conclusion (Parfit 1984, 388), (Parfit 2017).[2] Let us assume that the repugnant conclusion is inevitable both on Earth and in space, due to human reproductive dynamics, which, let us assume, will always lead to overpopulation (which is consistent with Parfit's highly simplified model that equates severe overpopulation with drastically deteriorating living conditions). If overpopulation is inevitable even in space, and even assuming multiple colonies and multiple planets settled by humans, would it still be worth trying to save our existence knowing that the repugnant conclusion is inevitable?

We might ask if we could have known before we were born that our only form of existence would be in a space colony with a low quality of life, would we have wanted to be born at all? And if we valued our existence above all else, what kind of sacrifices might we accept? Similar questions and thought experiments can also be posed in relation to human enhancement, asking, for example, what costs (side effects) we might agree to when there are good reasons to acquire a therapeutic or enhancement-only biomodification, complicating the terms of the thought experiment depending on whether the modification would be needed to survive or, for example, to realize a so-called life worth living, but with some side effect.

The risk of overpopulation and drastic deterioration of quality of life (repugnant conclusion) in a space colony—where the tipping point may come sooner than on Earth due to smaller living space and resource limitations and difficulties in resource production—is itself a serious challenge to the ethics of quality of life. The quality of life in space will be diminished regardless of population size due to specific factors of the space environment. In such a way, we may be reaching a paradoxical conclusion. While we can still accept that we will strive to save human life in the face of a global catastrophe and, to that end, evacuate the current generation to a space colony, the population ethic applied to space colonization should perhaps force us to abandon reproduction in space. Thus, space expedition would only be an effort to save the lives of currently living humans on Earth (probably only a small fraction of them), but no longer to save the species and ensure its continuation.

Only that such a conclusion would lead to an even more paradoxical conclusion. If we know that a given generation is surely the last generation of living people, what are our obligations to it? In other words, are our actions taken only because of duties to currently living humans, or because of duties to ensure the continuation of the species? Arguably, these two types of duties overlap because we assume that we cannot be the last generation. But if we are certain that we want to save the last generation of humans by colonizing space, it would be akin to persistently treating a patient who has no chance of survival. It would be a form of therapy at any cost, which will not work anyway (because the human species will end up with this last generation possibly sent into space).

In conclusion, the overpopulation problem may occur to an even greater extent in a future space colony.

Future People

Is it our duty to prevent the annihilation of humanity, or rather to support processes and phenomena aimed at the annihilation of humanity, and at least not to interfere with these processes, thus allowing them to occur on their own? For some, an example of the latter would be the current ignorance of climate change, which would be to maintain or even increase current levels of consumption despite the fact that it is leading to annihilation, and at least to environmental catastrophe. It seems that, without the alternative of space colonization, the rational choice remains the latter, assuming that climate change and overpopulation continue to increase, and that technological advances and economic development cannot offset the negative effects of these two increase. If we assume that, at least prima facie, moral duty is to fight suffering and human misery, and at least not to cause it, then we should be aiming for the annihilation of humanity (Leslie 1996, 173).

While this conclusion overlaps with the philosophy of antinatalism, it is rooted in a separate argument. For it proceeds not from an assertion of the fact of future human suffering, but from an identification of our moral obligation to prevent and minimize suffering. It may be, however, that the postulate of the annihilation of humanity is here derived too hastily, since a significant reduction in population may be a sufficient remedy. This depends, however, on whether the problem of human suffering is considered as related to the number of people or as a characteristic inherent in being human. In the former case, overpopulation is the problem, and it can be assumed that reducing the population to a certain level, with the technology, medicine, and infrastructure we currently possess, could provide a better quality of life for a smaller number of these human lives. I refer here to the famous Parfitian outcomes A and B, in which in one population we have a smaller number of people with a higher quality of life and in another a larger number of people with a lower quality of life (Parfit 1984), in which the correlation between number of people and quality of life changes dynamically in a Malthusian way.

In the second approach to the issue of suffering in the world, suffering is an inherent feature of being human and befalls even the wealthiest in the form of disease, depression, and finally death. Therefore, the concept of a moral obligation to reduce human suffering must clearly advocate one of these two approaches. Antinatalism is a total approach that qualifies every type of suffering as a disqualifying postulate to maintain the survival of humanity. I appeal here to the first understanding, where the misery characterizing humanity is related to the effect of overpopulation. In contrast, the inherent problem of our responsibilities to future generations is the aforementioned quality of life in space, which can be drastically low.

Whereas I took as my default assumption the intuitive conviction that human life must be protected as a species, and that the colonization of space under certain conditions is a rational and perhaps even necessary action to achieve this goal, here I pose the question of whether we have the right to do so because of the future rights and interests of future humans. In other words: do we have the moral right to save future human beings by creating the conditions for their existence?

Parfit's reflections on our relationship to future humans are useful here. Parfit considers a scenario in which our actions enable distant future generations to be brought into existence, but the effect of those actions will be a more rapid death of future generations. Their death will also occur as a result of our present actions.

In Parfit's example, it will be caused by the disposal of radioactive waste, which will be activated in the distant future and thus shorten the lives of future humans (Parfit 1984).

The dilemma that arises here is as follows. We can accept that it is better for future humans to exist at all than not to exist at all. Life is the basic good that makes all other goods possible. We are doing the right thing by giving future people a chance to come into existence so that they can benefit from such a good as life. On the other hand, I am not arguing that by not bringing future people into existence we do them harm, because we are not dealing here with a subject of sensation and consciousness (future people are not rights-holders). If we can bring future humans into existence being confident that we will create good living conditions for them, we can do so. From this point of view, in Parfit's example we are dealing with a situation where future generations, even if they die earlier due to environmental damage also caused by us, will experience more benefit than harm from us, because if it were not for our actions they would never have existed at all. However, our action of calling future generations into existence violates the obligation to uphold their rights, that is, the right to a long and good life, which, however, they cannot obtain because of the aforementioned environmental damage.

The example of Parfit shows that both solutions are both good and bad. Bringing future generations into existence with the knowledge that they will live a shorter life than we do—assuming, however, that this shorter life will be normal, without suffering—is good, because otherwise these people would never exist. On the other hand, our action is bad because we bring people into existence without respecting their right to a full and long life. Unless, as Parfit suggests, these future persons abrogate their rights and express satisfaction that they were called into existence knowing that otherwise they would never have existed even though their lives are shorter.

The distinctions introduced in Parfit's example are useful for considering our moral right to colonize space by planning the lives of distant future generations beyond Earth. Let us assume for the sake of argument that life in a space colony will not be able to be the same in quality of life as the average good life on Earth. Leaving aside the risk of constant danger in the form of a major technical malfunction, life support system failure or otherwise, let us assume that future generations will live much shorter lives in a space colony due to constant exposure to space radiation, which incidentally allows for a fairly close analogy with the Parfit example.

If we assume that calling future generations to exist under space colony conditions is doing them a disservice because we are unable to respect their right to live the same life as ours, then we should refrain from any action to prolong the existence of our species by means of space colonization. Such a decision would have far-reaching consequences because it could lead to the abandonment of the entire human space program. This is a rational option if we assume that the secondary goals of space exploration such as scientific mission and commercial exploitation can be satisfactorily accomplished by robotic missions, especially since the continuing potential risks to human health and life may not justify sending astronauts to missions of secondary status (Goldsmith and Rees 2022).

In conclusion, if we are confident that a colony in space will provide future generations with a life worth living, it is our duty to try to save the existence of our species—and at least increase its quality of life—by colonizing space even if the life expectancy of at least the first generations would be shorter than earlier generations living on Earth.

Antinatalism in the context of space colonization

Suppose we are utilitarians and our goal is to maximize happiness (Tännsjö 1998). According to the average view, we should care about the level of happiness felt by individuals in a population. This should be a life of a fairly high or decent standard, free from various sufferings and problems. Probably such a population should not be too large, because overpopulation tends to make the standard of living worse, although this is not the rule, because the ultimate effect also depends on the availability of resources. Thus, at least before reaching the point of overexploitation or superexploitation, where there is no return to a moderate policy of resource exploitation, such a population can be very large as long as there are enough resources, and the large size does not generate other problems such as crime or disease.

The alternative is the total view, where we can imagine a population that is incomparably more numerous than the current population of Earth, but the sum of individual happiness—even if the unit of happiness per individual member of that community is extremely low, incomparably less than the level of happiness in the average view—the sum of those small amounts of happiness ultimately exceeds the sum of happiness in the average model. It is hard to say what vision of the world seems ideal to us, and whether indeed the vision assumed by the total view offers a better world (de Lazari-Radek and Singer 2017, 114).

The various antinatalist arguments can be boiled down to one overarching idea. It would be better for humanity to cease to exist than for it to continue to exist and thus cause suffering to future generations. This is because even despite the undeniable pleasures experienced in life, net harms outweigh net benefits (Benatar 2006, 1). Negative utilitarians will add that what matters is pain, not pleasure (Belshaw 2021, 142). David Benatar convincingly points out that there is an asymmetry in the world between the pool of suffering and the pool of pleasure for the sake of suffering, which consequently, from the point of view of quality-of-life ethics, leads to the constatation of an insufficient average quality of human life (Benatar and Wasserman 2015).

Asymmetry implies that we have duties to prevent and at least not cause suffering, while we do not have analogous duties to cause or increase pleasure. This axiological asymmetry forms the basis for the asymmetry of procreative duties, which says that we have a duty not to bring into existence people who may have lives of poor quality, while we have no analogous duty to create people who may have good lives (Benatar 2015, 25). The problem with this approach, which in itself is not yet antinatalism but can provide an excellent starting point for it, is that we always experience some kind of suffering some of the time (and this is what we should prevent), and even if we experience moments of pleasure, we in turn have no reason to create lives that do not previously exist for those pleasurable moments (Belshaw 2021, 100).

However, the possibility of space colonization and human settlement opens up new possibilities for antinatalism. Let us imagine a situation in which, after surviving for a sufficiently long time and developing appropriate technologies, it will be possible to guarantee survival under conditions generally considered good for future generations in a space colony. Let us also assume that this population will be able to reproduce itself, and will therefore increase in numbers in defiance of antinatalist philosophy. For the proponents of human reproduction criticized by antinatalists, the issue here is the transition period during which humanity will develop engineering and medical sciences designed to guarantee an adequate standard of living for a continuously reproducing colony in space in the future. Thus, such a colony could reproduce itself, at least for a time, and transcend the objections formulated by the antinatalists about bringing more and more people into the world under increasingly deteriorating conditions.

While this objection may remain valid to the ever-expanding population on Earth, as noted, it may not, at least for a time, apply to a space colony population that has not yet reached a critical point that on Earth has long since been passed. Perhaps the population in space will never exceed such a point, not only because humanity may have learned its lesson from the population crisis on Earth. It may also be because humanity's growth in space will be a product of technological progress, which will open up ever new destinations and locations for human exploration and exploitation. Such a process would be a kind of misappropriation of Malthus's law because, at least for the not-so-large initial population of space colonists, systematic population growth would be correlated with a systematic increase in resources to be exploited. However, if some variant of Malthus's law were to become a reality in space as well, humanity could then enact a population policy of temporarily stopping reproduction and minimizing resource exploitation. This could be done through environmental rules such as the 1/8 rule, which assumes that the human population in space could only exploit a maximum of 1/8 of the available resources (Elvis and Milligan 2019).

It is worth adding here that the space environment offers raw materials with current exploitation potential, which, along with the development of technology, would open new locations in space. In this light, the space environment and the concept of humanity as a multiplanetary species does indeed open up new possibilities for rejecting the otherwise rational antinatalist postulate. But this requires the fulfillment of several technical conditions, which are not simple.

First, the degree of technology would have to reach a level of development qualitatively superior to the current technical possibilities of space travel and life. This is somewhat of a vicious circle argument, because if humanity does not reach this level, it will not be able to reproduce in space, so, somewhat paradoxically, the insufficient degree of development of space technology favors the antinatalists for the time being, because it makes any reproduction in space impossible.

Second, however, humanity would have to master the technologies of exploitation and utilization of raw materials in space, especially planetoid space mining. If we assume that the main context causing suffering on Earth is overpopulation, then this condition will certainly not be met immediately in the case of a newly created, self-sufficient space colony. The achievement of overpopulation understood as the necessity of a kind of struggle and strong competition for various resources, as a result of which a given population can be sorted into winners and losers, and thus provide rational arguments for antinatalists, may occur in a space colony, but it is not clear how quickly.

The antinatalist argument can also be undermined by applying genetic modifications to future astronauts and space colonists. These modifications are understood here as pragmatic, health-related modifications of a preventive or curative nature. Astronauts so enhanced—the purpose of such enhancement would be to make them better adapted to the conditions of altered gravity and cosmic radiation—could be free from many of the afflictions suffered by the inhabitants of Earth as a result of various diseases. Thus, such biomodification of future space colonists would also take away at least some of the justification for the rational philosophy of the antinatalists.

It is worth adding here that the concept of intergenerational care and justice cannot be attributed to the last Earth generation in relation to the first generation living in space. These postulates have force only between generations living on the same planet. Therefore, the last generation on Earth cannot be accused of having brought future humans into existence in—what we imagine to be unfavorable from our present earthly perspective—conditions in space. No future human being born in a space colony will certainly be able to compare the conditions of her life with the conditions of life on Earth from a first-person, subjective perspective. Therefore, it cannot be said that she was brought to live under objectively inferior conditions, because there is no first-person continuity to measure this. This will be even more impossible to establish, if future people born in a space colony were subjected to obligatory GGE.

Then first colonists, who were born on Earth and as adults left Earth in order to colonize, for example, Mars, also cannot input their first-person perspective comparing conditions of life on Earth with conditions of life on Mars with feelings of future people born on Mars. The latter, unlike the former, were not only born on Mars and never lived on Earth, but were also subjected to genetic modifications at the stage of embryonic development that partly at least made them different humans in the sense of their genetic adaptations. Thus, while we may intuitively assume that the living conditions on Mars for at least the first generation of future humans born there may be worse than the living conditions of the last generation on Earth, this objection cannot be applied to the first-person perspective of these future humans.

Moreover, it can be presumed that living conditions on Earth will have to deteriorate to a significantly large degree if establishing a colony in space, under conditions intuitively often perceived as more difficult, becomes the goal of humanity. In this sense, the plea from the diminished quality of life in the space colony can be considered self-refuting as long as, by the planners and organizers of the space colony, evacuation to space is considered more attractive than staying on Earth.

The arguments of antinatalists, which are rational in the terrestrial environment, will therefore, apparently, be weakened in the space environment by the fact that, due to progressive overpopulation on Earth and environmental destruction and climate change, life in a space colony will, by definition, always be better than living conditions on Earth. We might add—at least up to a point where the population in space does not reach overpopulation. Every future person born on Mars is born into conditions better than those of their predecessors, especially previous generations living on Earth, or generations simultaneously living on Earth. Thus, the argument of those antinatalists who appeal to intergenerational justice and care is undermined here.

In a space colony, the following scenario is also possible, which can come true when a generation of space colonists decides that it makes no sense to continue humanity's life. Their moral right not to undertake reproduction will be reinforced when they ascertain that there is also no chance of improving the lives of humanity living in a space colony. Then this generation of colonists would perhaps come into possession even of the right to annihilate themselves.

It can be said that a generation of space colonists may hold some kind of beliefs appropriate to philanthropic antinatalism, with the antinatalism being much more radical and justified than the antinatalism shared on Earth. Let us call them cosmic antinatalists. On Earth we have a rational basis for believing that life for future generations will be worse because of overpopulation and climate change, which provides a legitimate basis for philanthropic antinatalism (Brown and Keefer 2020). In contrast, recognizing the hopelessness of human life in the cosmos will provide the ultimate basis for antinatalism.

Finally, according to Benatar's antinatalism, a program of space expansion to prevent the extinction of humanity contradicts Benatar's conception of humanity's extinction happening sooner rather than later (Benatar 2006, 194). Postponing it through space colonization measures only prolongs the suffering experienced by humans, and the point, after all, is not to bring into the world—whether on Earth or in a space colony—new generations that will suffer between the current generations and the last one (Benatar 2006, 199). This is all the more possible because, at least from the perspective of today's technical capabilities as well as the advancement of medicine in space, missions even to Mars, such as the Mars One, a one-way project, are treated by some in terms of heroic and even suicidal missions (Arnould 2017, 35).

Since the optimal population is zero (Benatar 2006), the extinction of humanity is not what we should fear, and the postulate of saving them through space colonization becomes devoid of justification especially in light of the fact that concerns about experiencing suffering in space are very legitimate.

It is worth adding that the political context will serve to reinforce rather than reduce the level of risk and danger in space. Space exploration and exploitation itself is usually associated as a new theater of the political and even warfare (Szocik et al. 2017), militarization of space becomes a reality (Steer and Hersch 2021, 301), and at the very least the notion of an arms race should be taken realistically (Grego 2021, 265–267). The context of annihilation of humanity and colonization of space to protect at least part of the population will probably lead to conflict between space powers and a struggle for space in a space colony.

Another problem, however also related to social and political dynamics, is how to organize the space colony politically (Wójtowicz and Szocik 2021). There are strong reasons to believe that the sacrifice of certain values in the short term, especially values concerning democracy and freedom, will be required to achieve the desired long-term stability (Schmidt and Bohacek 2021). These, then, are the factors that will diminish the quality of life and make the lives of future people perhaps unworthy of further life, which in turn calls into question the sense of concern for making the existence of future generations possible, in an antinatalist sense.

As the scenario I am considering extends into the future, it can be assumed that radical human enhancements will be applied broadly, not excluding GGE as well, and that today's distinctions between therapy and enhancement and between health-related and non-health-related modifications will perhaps have no relevance to the everyday application of enhancements. It is a variable that can increase the quality of life and make the lives of future generations more worth living. Genetic modifications will perhaps eliminate diseases, making life more pleasant and comfortable (Buchanan et al. 2000, 1). Consequently, a geometric increase in population, whether on Earth or in a space colony, need not at all lead to a decrease in the quality of life for future humans, and thus may weaken the arguments of antinatalism advocates.

In sum, if we take the rational but counterintuitive philosophy of antinatalism seriously, we find that the idea of future colonization of space to save our species is precisely what we should avoid at all costs.

Can Any Others Have Any Rights as to Future Persons?

The discussed problem of future generations and population ethics leads to the following paradox. Humanity aware of the challenges discussed, such as those signaled by Parfit, can take appropriate collective action in advance to design the future. But this designing of the future does not have to be done by conventional means alone. Biomedical technologies with widely applicable GGE are and will be available in the future. The paradox is that awareness of a problem in the future and a morally good intention to avoid the problem may lead some, and humanity simply, to interfere with reproductive rights, autonomy, and freedom.

It is something rational and logically conclusive to recognize, at least in theory, that the better the qualities and characteristics of individuals, the better the quality and well-being of society as such. And thus, it is therefore in the interest of society to have as many of the best possible individuals as possible (Sparrow 2016, 133). However, unlike the abstract principle expressed by the parents and the virtual right of the future child to well-being, this rule transferred to society is accompanied by special problems that those two levels do not have. One of these is the establishment of a body which would have the right to decide, and at least to suggest or recommend, any genetic modification of embryos dictated precisely by the good of the community. Who would constitute such a body?

The collective action problem recurs here, which, however arguably harmful to the population, somewhat paradoxically serves to protect individual rights and, at least for the time being, protects against the aforementioned risk of official, institutional dictation of the future of individuals, especially children. Prospective parents are not likely to be guided by the variously understood good of society. On the other hand, such a scenario may occur in the case of GGE applied to space missions, both on Earth and during reproduction in space.

As Thomas Douglas notes, the concept of recognizing some third-party rights over how a given child develops and is raised could theoretically justify the demand for appropriate genetic selection of the embryo toward the elimination of those traits that will pose a problem for society (Douglas 2019, 309–310). Douglas presents a much stronger position than mine, because I am considering a critical situation from the point of view of humanity, in which the survival of the species *Homo sapiens* is threatened. Douglas, on the other hand, considers a normal situation in which parents should already raise their children in a way that does not harm society. Instead, he adds a new tool to these measures, genetic selection.

The costs, primarily economic, associated with the birth of a new child and her subsequent upbringing are, to some extent, borne by society as a whole. This fact may entitle actors other than parents alone to—at least theoretically—influence reproductive decisions. Society participates not only in the costs but also in the benefits that raising children into future adults will bring to that society and to humanity as a whole, and therefore, at least virtually, is interested in making the best kids possible (Anomaly 2020, 74). For this reason, succession of generations implies the bearing of costs as a necessary toll that present generations pay in order to make themselves viable in the future through each new generation, which requires great resources in the process of upbringing.

In the context of future space missions, in the hypothetical scenario of shaping embryos for space missions (whether on Earth or in a space colony), the context of the missions, as well as their rationale and purpose, can at least hypothetically justify a moral right to embryonic modification to an even greater extent than for terrestrial purposes. Every type of space mission, including scientific and commercial missions, justifies such a procedure with the principle of beneficence and reciprocity that guides a society that invests in the upbringing of children. It is an intergenerational version of return on investment. Today's generation, from the point of view of its own interest (the selfish premise) and the interest of the whole species (the altruistic premise, but rooted in the previous selfish premise), should modify the embryos destined to participate in space missions, because only in this way can it be certain that the costs incurred in raising future children sent on space missions will certainly pay off, or at least significantly increase the chances of such a return.

In this case, the nature of the space mission is irrelevant. What is relevant is the return on investment associated with the fact that the reproductive rights of parents are always associated with numerous obligations and costs that society must bear in order for parents to exercise their reproductive rights. Thus, in this approach, it does not matter whether the mission objective is scientific research (least benefit to humanity), commercial purposes (great potential for humanity, but arguably the possibility of a capitalist course of space exploitation is higher than prosocial exploitation for the benefit of humanity), or space settlement founding (greatest benefit to humanity). What justifies the modification of an embryo here is not the expected effect of the mission, but the costs incurred in raising an already modified human being before birth.

In conclusion, while the idea of any third party planning the specifics of future children on the GGE path in a space colony is a dystopia that we should avoid, humanity in the future should consider all possible scenarios in which such a policy would have only positive effects and is necessary for the survival of our species.

Do We Have a Moral Right to Apply GGE to Space Exploration?

The argument from an open future takes on particular significance in a situation where GGE would be applied on Earth with the goal of preparing children to participate in a future space program. This form of biopolitics is possible in a situation in which life in a space colony would be widely perceived as a promising lifestyle, guaranteeing better conditions and prosperity than life on Earth. I also assume here that the changes made under the GGE will impede life on Earth after adulthood, while they should not impede that life before readiness to participate in the space program. This issue is very controversial primarily because children so modified will continue to be born on Earth. And it is difficult to say whether a modified child will not want to remain on Earth. I assume in this scenario that life on Earth will be possible and that there will be no compulsion to migrate to colonies in space.

To some extent, the analogy with economic migration on Earth may be useful here. Imagine that residents of country A often migrate upon reaching adulthood to country B. Life in country B is widely viewed in country A as a dream place. However, not all residents of country A decide to emigrate despite the fact that they can do so and that life in country B would actually work out more favorably for them than staying in their home country. For some reasons, these people decide to stay in country A.

It cannot be ruled out that similar will be the structure of the future population of Earth, which perhaps even in majority would prefer to migrate to planet B or just any available space colony, but for reasons analogous to those guiding the decisions of migrants on Earth will give up such migration. We are then in a situation where we know that most children born on Earth will want to migrate to a space colony, but at the same time we know that a certain proportion will certainly choose to remain on Earth. To modify the latter group, therefore, would be an injustice to them, just as not modifying the former would be an injustice and neglect to the former group.

If we adopt a utilitarian perspective, we might conclude that in such a situation, in which we do not know which embryo should be modified to equip it with adaptive traits for life in space, we should obligatorily modify all embryos based on this assumption that the majority of individuals will want to leave Earth. This is acceptable on utilitarian grounds, but it also introduces a certain component of paternalism, in which the state or some other body determines the lives of future humans. This seems to be a maximum that could be accepted under certain conditions. Perhaps some would accept a more radical position, which would derive from the fact that each embryo is preemptively modified an obligation for each modified embryo to colonize the cosmos. In such a situation, perhaps not every embryo would be modified, but only as many as would be needed to relieve population problems on Earth and to maintain colonies in space. However, this would then be unfair to both the modified children (because not every one might want to be forcibly sent into space) and the unmodified children (because some of them might want to fly to the space colony, but the lack of GGE prevents them from doing so).

Finally, perhaps there would be even more radical proponents of forced colonization and forced GGE who would find that the solution to the problem would be mandatory GGE for all supplemented by moral or psychological GGE. The purpose of such moral and psychological embryonic modification would be to design the mentality of the future child either in such a direction that it would want to fly into space itself, or in such a direction that it would be docile and amenable to all persuasive and educational influences.

The very notion of an open future and the postulate of just leaving it open is not possible here. Both the decision to pursue GGE and the decision to abandon it for a particular embryo is a decision that determines that embryo to pursue a particular version of the future, with no possibility of changing their fate. We can gradate and manipulate the terms of this thought experiment, starting with a scenario in which life on Earth is most bearable compared to life in a space colony, to a variant in which life on Earth is still possible but extremely oppressive. Then, using at least the reflective equilibrium method, we can test our moral intuitions and confront them with moral principles and ethical theories. We can assume that our intuitions may tell us that in the first boundary case, mandatory modification of all embryos is far from prudent, while in the second boundary case, it is worthy of consideration. But we still know that in both cases there may be many exceptions—future humans who will not accept the fate prepared for them. The numerous intermediate stages arguably further complicate the decision, as they introduce more room for exceptions to both options.

The case discussed here differs significantly from analogous cases discussed by philosophers with respect to imaginary scenarios on Earth. These philosophers typically analyze scenarios involving more or less trivial genetic determinations, such as whether to program a future child with genes that make them a musician or a poet, a mathematician or an athlete. These examples are thus about enhancement for enhancement's sake, rather than referring to the context of environmental determination and related to the health of the modified child, as in the case of space. With respect to the space environment, the discussion of the open future and GGE is not about what kind of work the child will do, but where the child will live—on Earth or in space.

Reproduction in a future space colony—assuming that from a biological point of view reproduction in space will not be a problem—is not necessary on the assumption that either (1) the human population can continue to exist on Earth where it can reproduce, or (2) the human population cannot continue to exist on Earth, but we do not care to continue the existence of our species through reproduction beyond saving the last generation by evacuating to a space base. The latter case is a form of antinatalist philosophy and can be justified for a variety of reasons.

With respect to premise one, if humanity has to construct a new human settlement from scratch anyway, according perhaps to new laws and principles, space colonists will not be forced to replicate the regularities known from Earth, where in principle almost all countries value increasing their numbers through reproduction more highly than through immigration. A space colony, therefore, need not increase its numbers by breeding within the colony, but by migrating from Earth (assuming that such migrations are technically and economically possible).

We may be faced with a situation in which individual reproductive freedom and autonomy are severely limited and even suspended. It is worth distinguishing between two contexts here, that of health, which might abolish this autonomy, and that of nonhealth, which, if it did abolish the reproductive autonomy of future parents, would limit it to a lesser degree. The vision of the future in space is thus the product of a difficult dilemma between the desire to save the species—for I assume a degree of optimistic faith in humanity, and believe that it can only lead to the creation of a somewhat authoritarian space colony if that is the only form of humanity's existence currently available—and strong antinatalist, and virtue ethics-rooted, arguments that it is better not to continue the existence of our species under certain conditions that make our lives barely worth living (and perhaps not even worth living).

It would be good if every space colony could guarantee human rights, the right to migrate, the right to return to Earth (if returning to Earth is a value for any reason), and many other rights that we value on Earth today (Mazur et al. 2020), (Schwartz 2020). But I recognize that, for environmental and technological reasons, perhaps humanity will face a dilemma: choose a life that is barely worth living (another question is whether humanity will be aware that it has set in motion a sequence of events leading to a life that is barely worth living, or will only find out after the fact), or accept extinction.

9
Conclusions

Imagine humanity one hundred years from now. Will the average quality of life on Earth then be worse or better than the average quality of life today? Will we have an advanced space program and space will be a natural extension of human scientific activity, primarily commercial, and perhaps even begin to settle it? We do not know, because demographic projections indicating increasing overpopulation and climate change (which may reduce quality of life) may be offset by advances in science and technology (which may either keep quality of life the same or increase it). Instead, we may wonder whether a world in which humanity pursues long-duration deep-space missions might be a better world than a world in which space exploration stops at its current orbital level.

As I pointed out in the book, there are many factors and many unknowns that require us to consider different scenarios. What interested me was the bioethical context of the future of humans in space. Fundamentally, I believe that we should look to our future with openness and hope. Even if human enhancement understood in the way discussed in this book turns out to be necessary.

I believe that as humanity we should not slow down development in science, including biotechnologies, for reasons other than safety. It can be extremely difficult and lengthy to get a 100 percent safety guarantee on gene editing, especially GGE. But as long as we can prove the safety of this procedure, there are many strong rationales for applying gene editing for a variety of purposes. Especially since, in certain scenarios that I have considered, it may be necessary for allowing us to continue as a species.

I also believe that it is worthwhile to change the approach in thinking about the future of humanity in applying new technologies from one based on fear, distrust, and skepticism to one based on trust, hope, and positive thinking. Skepticism about gene editing perhaps causes damage similar to what opposition to in vitro fertilization, transplantation, and other medical procedures—where that opposition exists—has caused and is causing.

One hopes that, ultimately, humanity will simply approach the issue of applying biomodification pragmatically and consequentialistically. I believe that space exploration will open up so many possibilities, and provide so much incentive for long-term missions, that if at some point humanity decides that these goals in space cannot be achieved without human enhancement, it will decide to apply it, however cautiously.

The Bioethics of Space Exploration. Konrad Szocik, Oxford University Press. © Oxford University Press 2023.
DOI: 10.1093/oso/9780197628478.003.0009

Bioethics of Space Missions in the Light of Futures Studies

Imagine life in a future settlement in space, such as Mars, which is life worth living. Such life should be defined in comparison to the quality of life on Earth. Let us assume that it will be a quality of life at least equal to the quality of life on Earth at the same time. What could argue against life beyond Earth if life in space were not only not inferior to the quality of life on Earth, but, moreover, could be characterized by greater stability and lesser degree of occurrence of various types of existential risks?

One can add to this scenario, entering somewhat utopian ground, that perhaps only life in a space colony would have a chance of realizing an idyllic scenario in which the ideas and principles of justice, equality, peace, and cooperation would be implemented from the very beginning of the extraterrestrial population. If life with a high quality of life in space is possible, and will require human enhancement as a prerequisite, even by way of germline gene editing (GGE), why not open such a possible path of development for humanity? Especially when it may enhance, and perhaps be the only way to guarantee, the chances for the continued survival of the species.[1]

I began chapter 1 with the proposition of imagining a negative scenario that we would want to avoid. Bioethics is about black scenarios, although it is not usually applied to futures studies and addresses present-day biomedical challenges and problems on the horizon. Nevertheless, bioethics focuses on finding threats and risks. Futures studies in my approach, and by extension bioethics as applied to futures studies, tends to focus on outlining preferred scenarios. Futures studies also discusses possible and probable futures as well as preferred ones (Bell 1996a, 81). It is hard to talk about preferred scenarios if we do not consider worst-case ones. And rejecting the concept of human enhancement for the sake of the various arguments I consider in chapter 3, which is a possible or even probable scenario, can lead to a worst-case scenario.

Preferred futures are those in which humanity will have lives of the most worth living possible given the possibility of expansion into space. This means both humanity living on Earth and carrying out space missions for scientific and commercial purposes, and a more distant, catastrophic scenario in which humanity will be forced for existential reasons to settle beyond Earth, and Earth will either be uninhabitable or remaining on Earth will guarantee a life far less worth living than life in a space colony. I discuss scenarios that assume that the preferred futures for humanity are those that require a permanent human presence in space. Such a presence could replace life on Earth as impossible or difficult after a catastrophe, or complement it, enhancing well-being to a degree not possible on Earth. The presence in space can solve population and environmental problems on Earth, can bring scientific breakthroughs applicable on Earth, and open up new areas of development and inspiration that we are not even aware of today.

While these preferred futures are possible, as well as probable, there are many factors that may cause them not to be realized. I also consider worst-case scenarios. These worst-case scenarios include both futures in which we stay on Earth and forgo long-term and deep-space exploration, and those in which we choose to do so but which worsen our well-being. At the same time, they may worsen the quality of life of either the population living in space, or just the population remaining on Earth, or both.

As much as I am inclined to argue that the preferred futures for humanity require space exploration rather than just staying on Earth, I am aware that in pursuing these futures, at some point something can go wrong. And it will not necessarily be the bad intention of some individuals, but a confluence of events reinforced by the specific harsh conditions of life in space. This worse-case scenario of the future in space is a situation in which, against our will, we generate, as humanity, such social and political, but also infrastructural structures in which we deprive ourselves of basic rights and freedoms.

These worst-case scenarios can lead to an existential catastrophe. It is worth recalling here the classification of catastrophes proposed by Toby Ord, in which dystopian futures are also a type of catastrophe. Ord points out that one kind of existential catastrophe, along with extinction (i.e., future scenarios "without humans") and "unrecoverable collapse of civilizations," are dystopias, that is, future worlds with a civilization that is admittedly bad, completely valueless, or of minimal value. Ord divides dystopias into undesired (where people do not want such a world, but the structure of society determines fate against people's will), desired (where people want such a world, but the world does not meet their expectations), and enforced (where only a small group wants such a world and forces the others against their will) (Ord 2020, 153–154).

The type of dystopia that can be realized in the cosmos, but that we as humanity want to avoid, are desired dystopias. It is worth emphasizing, however, that all types of dystopias can be realized on Earth, and there is no ontological feature or characteristic in space that makes space settlement more prone to the realization of any of the dystopian scenarios mentioned. Sometimes philosophers of space exploration like to talk about the potential dystopian threats supposedly particularly generated by the concept of space colonization. This would be due to harsh living conditions, impossibility or difficulty of migration, restriction of certain rights and freedoms for infrastructural reasons, and life support system. It is worth remembering, however, that the dystopias realized so far in human history were realized under conditions in which the possibility of escape, however difficult it may be, was nevertheless easier than what we can now only imagine about the possibility of migration in a future space colony. Restrictions on access to food and medicine and medical care were also a big problem in these Earth dystopias, but nevertheless it was a different level of problem than the analogous problems in a space colony. Thus, it is not so much the infrastructural constraints and the harshness of the environment that generate the risk of dystopia as it is the bad people wielding power, as human history has shown to be free of these specific difficulties and constraints that characterize the space environment.

I see bioethics as a more restrictive approach that, as I mentioned, exposes risks and threats rather than opportunities and possibilities. Supplementing this precautionary approach of bioethics with futures studies, in turn, ensures that these negative scenarios in the cosmos cannot come true. Dystopias are a self-refuting possibility because futures studies are concerned with considering those preferred futures that will always meet minimal ethical requirements. These requirements are to guarantee freedom and welfare for future generations.[2] Whether on Earth or in space, futures studies is a kind of heuristic that ensures that always these two values must be guaranteed.[3] Thus, considering negative scenarios is only meant to make us aware that at some point in the future, choosing this trajectory rather than another may bring negative consequences, so we should either avoid choosing it or look for countermeasures. In a sense, the pessimism of bioethics is balanced by the optimism of futures studies. Futures studies are not simply a consideration of the future, but only of futures that provide freedom, welfare, and respect.

Wendell Bell emphasizes that the futurist has a duty to care for the good future of humanity, and especially to express the interests of future generations. From this point of view, the idea of futures studies excludes the possibility of allowing scenarios that may limit the freedoms and welfare of people in the future. As Bell points out, regardless of preferences, methodologies, and preferred future scenarios, the task of futurists is to analyze and support the pursuit of only those scenarios that guarantee the protection of these values (Bell 1996b, 158). In this sense, the objections or concerns of, for example, Schwartz (2020) and Green (2021) that I mention in the book about the possibility of developing a totalitarian scenario in space also in the context of human enhancement have no basis because they contradict the very idea of futures studies as analyzing the future in moral terms. The analogy to Asimov's laws of robots is useful here. Just as robots can never, according to Asimov's concept of rights, do harm to humans, for then they would abolish themselves and contradict themselves (Asimov 1956), so there can be no futures studies without a moral component and the protection of the interests of future generations.

Many arguments can be made for the obligation to care for future generations, but one argument seems particularly compelling. It is the thought experiment invoking the veil of ignorance, where we make a decision to choose a future scenario in a situation where we will be randomly placed in one of the scenarios but do not know which one (Bell 1996b, 300). Therefore, according to the concept of rationality, we should always and everywhere choose such models of development that each of them guarantees the highest possible quality of life for future generations.

The aforementioned fact of looking at bioethics through the prism of futures studies—as it cannot be otherwise if we are talking about the possibility of applying human enhancement to space missions yet to be realized in the future—makes me present even radical forms of human enhancement as scenarios that may be preferred and even desirable, at least under selected circumstances in selected future scenarios.

Much depends on the quality of life on Earth, and whether Earth maintains its position as the best place for humans to live. The fact that it is the only astronomical object to which we are evolutionarily adapted does not preclude the possibility that—despite the lack of this historical evolutionary adaptation to life on other objects in space—life in space may offer a better quality under certain conditions. It may be more stable in the sense of maintaining the continuity of a stable civilization, which on Earth at some point in the future may be shaken. It may be safer and more attractive in a social sense in terms of guaranteeing basic rights. It is a certain paradox of the space colony that, on the one hand, it can limit rights for technical and ergonomic reasons, but on the other hand, it can, as a new social project of humanity, be built from the ground up on the ideal and principles of equality and justice. Life in a space colony may be particularly attractive to those who are born there and have no comparison with life on Earth.

But even if the future confirms that only expansion into space can guarantee preferred futures for humanity, it may involve a first-generation problem. This is the type of problem that could inhibit human deep-space missions as well as the application of new biomedical technologies, with GGE at the forefront.

Let us conduct the following thought experiment. Imagine that we have a time machine that can transport us to the future.[4] What we know is that this will be a future in which humanity will inhabit some astronomical object. Knowing this today, would we want to travel in time to that space colony of the future? And if so, would anyone choose to move to this colony without the possibility of returning to Earth to the present day (believing that life will be more worth living in space in the future)? Would those who chose to relocate to a space colony do so only on the condition of being able to return to Earth?

Finally, imagine that we are given the opportunity to move back in time, but only to play the role of pioneers. We know that our agreement in the pioneering mission is necessary for the possibility of building a guaranteed welfare future space colony. The purpose of such a mission is to save humanity, and without us, as the first generation to carry out such a mission, it will be impossible to accomplish the mission and, by extension, to save humanity.

Brian Patrick Green points out that the backup Earth concept offers only two possible scenarios. Each is associated with risk, but each balances the effects of short-term and long-term risk in a different way. If we choose a future in which we do not pursue space missions to avoid exposure to astronaut death in the short term, then we increase the risk to humanity as a whole in the long term. If, however, we accept a fairly large short-term risk to astronauts, then we reduce the risk of species extinction in the long run, or prevent it altogether (Green 2021, 47).

Although my book is a book on the bioethics of space missions and is part of the philosophy of space missions, it should also be considered within the framework of futures studies. Such a perspective explains why we should consider future scenarios, and—even if not yet a real problem today—the future bioethical challenges in space that I discuss may pose a serious challenge. A futures studies perspective allows us to go beyond the here and now not only to prepare for the possible scenarios of our decisions today and in the near future but also to be able to plan for them, to eliminate randomness, and above all to realize the long-term consequences of our present actions.[5] The following definition of futures studies is particularly useful:

> The exploration of possible futures includes trying to look at the present in new and different ways, often deliberately breaking out of the straitjacket of conventional, orthodox, or traditional thinking and taking unusual, even unpopular, perspectives. It involves creative and lateral thinking in order to see realities to which others are blind. It involves thinking of present problems as opportunities and present obstacles and limitations as transcendable. It involves, most of all, expanding human choice. (Bell 1996a, 75–76)

The problem is how we treat what is new and what we interpret as an obstacle and barrier. Paradoxically, it is not always what we regard as an obstacle that is actually an obstacle, but rather our thinking and imagining about a given thing regarded as an obstacle, and our opinion and moral evaluation of a given type of future, that can be the greatest obstacle to its realization.

As Bell notes, people make two basic types of mistakes in thinking about the future. One is that we consider as possible what is impossible. The other, conversely, is to consider as impossible what is possible. According to Bell, the second type of error leads to worse consequences because it limits our options and narrows the horizon of future choices (Bell 1996a, 78–79). We should test our ideas and beliefs, and only when we see that we cannot realize them should we either look for new ways to realize them, including postponing them until the future while waiting for new technical solutions, or, when we are sure that they constitute an impossible barrier to be crossed, we should give them up.

With regard to the concept of long-term deep-space missions requiring the application of human enhancement, it would be better if we made the first type of mistake, which is to consider the impossible as possible. Somewhat paradoxically, this is the type of mistake we cannot make with respect to space missions, because both deep-space missions and human enhancement are merely more advanced and complexity tasks on a continuum of tasks and phenomena that we as humanity already know and are already accomplishing. We are talking here about future scenarios that will be new more quantitatively than qualitatively. Life in a base and even a colony on Mars could be considered qualitatively new and considered impossible before Yuri Gagarin and the Apollo program, not to mention the long history of space stations.[6] Today it is merely a continuation of that first step, probably considered impossible by many, that Gagarin took. We cannot make the mistake in this case of considering the impossible as possible, because it is already being accomplished. However, we can make the mistake of considering the possible as impossible and hinder our development, make it impossible to achieve certain positive effects for humanity in the form of increasing the quality of life or, ultimately, increasing the probability of saving humanity.

We should therefore modify our thinking about possible and probable scenarios in the future by keeping in mind that even the most probable future—and at least the future scenario considered as such from today's perspective—is itself contingent, because it depends on a number of conditions that must exist for it to be realized (Bell 1996a, 81).

In the same way, we cannot exclude the occurrence of a scenario considered less probable, such as the realization of space missions requiring human enhancement, because we do not know what conditions will exist in the next decades. The conditions inherent in the present may change to such an extent that maintaining the status quo will no longer be possible.

Thinking about the future, about preferred scenarios, does not have to be seen at all in terms of science-fiction, let alone utopia. The point is to think about positive images of the future. Bell cites examples from the past when considering future idealistic images in the Sumerian era, ancient Greece, the period of dominance of the Judeo-Christian religion, and finally the Enlightenment brought about positive, breakthrough changes (Bell 1996a, 85).

In relation to space bioethics and the philosophy of space exploration, we should, by analogy, consider idealistic images of our future in space, consider how to build it from the ground up to increase the quality of life of humanity not only in space but also on Earth, as well as try to predict and prevent undesirable directions of development. The future is unpredictable, but to some extent it can be shaped by our conscious, purposeful decisions and actions. Making a decision today to strive to make our species a species ultimately inhabiting more than just one planet may affect changes in other areas of our activity, including bioethics, if it turns out that an important element leading to the achievement of our goal, namely, expansion in space, is the application of human enhancements. This entails changes in our ethics and morality.

If recognizing the right moment to begin a sequence of events leading to preferred futures, or to constitute another watershed in that already begun sequence, will require changes in our morality, then we have a strong case for doing so. Such a turning point in our thinking about our future in space may be seeing the need for gene editing, considered controversial. Thus, we must make some decision about the trajectory of our future to accelerate and facilitate later events leading to preferred futures, and eliminate those that may later be difficult to eliminate or will be recognized as impossible to eliminate (perhaps erroneously). Therefore, as Bell rightly points out, thinking about the future is indispensable to our understanding of the present, our decision-making and action in the present, and our appropriate management of resources now and in the future (Bell 1996a, 90).

To this end, we not only can but should make predictions[7] considering as many variants of possible trajectories as possible, taking into account different conditions and different contexts that may happen in the future (Bell 1996a, 107). A single variable can rule out the realization of a given scenario.

Imagine if today's moral conservatism in public policy is maintained and GGE ban is not lifted. We do not even know what opportunities and trajectories for humanity as a whole this ban will deprive us of if it is maintained for decades to come. It may also inhibit human colonization of space. That is why I consider different scenarios, wondering whether a future in which we allow the use of even radical forms of human enhancement, and one in which we decide to develop without them, will be better for humanity. However, it is worth remembering that the latter option reduces, not increases, the number of possibilities.

Futures studies have some key assumptions.[8] Some of these are particularly important, such as the assumption that not everything that happens in the future necessarily includes events that have already happened or are happening now. This is the so-called possible singularity of the future assumption. Not everything can be predicted, however we can move in a certain direction. New events in the future may be without any connection to what we have now or what has happened in the future (Bell 1996a, 141–142). The future may take us completely by surprise and take us on a completely new course.

Consider the following example. Today we make various moral calculations when we evaluate the justification for applying biomedical technology. Often we dismiss them by saying, for example, that there are nonbiomedical alternatives. With regard to long-term space missions we might say that we do not need to consider radical human enhancement for settlement on Mars at all if we can build a shelter under the Martian surface to protect ourselves from cosmic radiation. But we can only say that today, only in recognition of that particular moral perspective that we have today, having only that knowledge and that experience that humanity has today. What we do not know is whether humanity in the future will not have strong reasons to live on the surface with genome editing instead of building a shelter a few meters below the surface of Mars. Leaving aside the fact that life on the surface is usually better for humans than under the surface, human life on Mars will be a new event in human history to such an extent that we should remain skeptical about the idea of applying our moral calculations and problem-solving approaches today to what will happen on another planet in the future. This is also the problem of the sense and validity of applying the moral principles and norms currently used on Earth to a new environment, where the challenge is not so much its novelty, but rather the impossibility of predicting whether what is a moral controversy for us today will remain so in the future.

Another key assumption of futures studies, the most useful knowledge assumption, requires us to refrain from the habit of uncritically applying old solutions to future problems. This assumption corresponds to the assumption about the unprecedented nature of future phenomena. As Bell notes, it may be that even if old ways of solving problems continue to work, new ways may emerge that are more reasonable. It may also be that the future will bring new problems requiring new ways of solving them (Bell 1996a, 146).

We should not keep planning to solve new future problems with the same old solutions, because we may invent better ways to solve problems in the future. We should not close our horizon of possibilities. Old methods may not fit new problems. With respect to bioethical principles and rules, it is hard to assume that we will invent new ones. Instead, what the future may bring is our shift either toward liberalization or conservatism, at least depending on context, place, and time. We can envision a situation in which, as with military ethics today on Earth, we will have two ethical systems, a liberal one in space and a conservative one on Earth when it comes to human enhancement.

Reference to a case study may be useful here. Often we do not have cases similar to anything that will happen in the future, so why should we assume that only old methods known from Earth can fit our future in space? Why should we assume that we can only protect astronauts with old-fashioned, conventional ways the way we've been used to protecting workers in harsh conditions on Earth for hundreds of years—with proper suits, properly thick walls, shelters, and so on. Why take this problem-solving thinking, which dates back to antiquity—and at least as long as humanity has been making armor—to our future habitat in space? Why not accept that new problems may require new solutions? Such a new solution is genome editing taking even GGE into account so where it can work perfectly well, rather than conventional countermeasures.

Adopting such a perspective requires modification of morality. But humanity has made such changes at the population level, not to mention the individual level. DeGrazia and Millum give as examples of such radical changes in our moral thinking our moral views about women and other races. They also suggest that other major changes in morality, such as toward animals, may await us in the future (DeGrazia and Millum 2021, 27). Why then should we not imagine other changes in our moral thinking, this time concerning the acceptability of biomedical technologies?

This brings us to the next two key assumptions of futures studies, namely the assumption of an open future and the correlating assumption that "humans make themselves." Humans are unable to transcend certain limitations, such as the biological limitations that only women bear children and not men, as well as the fact that all people will die someday. But these limitations are insurmountable as long as we do not decide to overcome them with technology. At least theoretically, men could bear children, as well as at least theoretically, we could someday significantly extend human life. Nevertheless, there are many limitations that are not insurmountable, and the future, as Bell points out, is not entirely predetermined (Bell 1996a, 150, 152). The task of futures studies is to constantly monitor the degree of openness of our future with respect to various variables and conditions, and to try to determine which of the observed constraints are insurmountable and which may be only temporarily insurmountable (Bell 1996a, 154).

It is worth recalling here the example proposed by Martin J. Rees. Rees points to space flight technology as an example of a technical constraint on the idea of space colonization. Rees says, giving an analogy to airplane flights, that space flight technology is like having to rebuild the airplane every time after every flight (Rees 2003, 173). Rees's example is a good illustration of the problem of different kinds of constraints, in this particular case technological constraints, which narrow our perspective on the future. Cognitive constraints lead to a situation in which we de facto assume that the future must remain just like the present. Or we can transcend some technological constraints that are constraints today, but will disappear in the future? Moreover, and perhaps more importantly, because it is the sine qua non for overcoming at least some of the technological and environmental constraints, is overcoming our moral constraints that cause us to often analyze new biomedical issues in the terms of a static, conservative society with only conventional medicine.

The future is open. As such, it remains uncertain. There are many alternative possible futures, and we can choose one particular trajectory of development. The future is thus not, as Bell points out, inevitable (Bell 1996a, 154). Because futures are open-ended and not inevitable, I want to consider whether one possible scenario, the implementation of long-term space missions using genome editing, might be a good scenario, and whether it will result in a worse future. And it is a possible scenario, given our plans for expansion into space, advances in science and in genetics. Should we fear and prevent this scenario? Or maybe we should already be making such decisions and taking such actions that human enhancement by gene editing will be used universally, and the context of space missions will be its specific, but not exclusive—however perhaps first used on a large scale—application.

And this brings us to the most important of the key assumptions of futures studies, the assumption that "some futures are better than others." We seek a better future, and we basically have a consensus about what a good, exemplary society looks like and should look like (Bell 1996a, 157). We desire a future that guarantees us a good quality of life, a life worth living, not its opposite. I relate this paradigm to the future in space—it is about the search for a future good society in space, and what might threaten it, that my book is about.

The tool used by futurists is telling stories to present alternative scenarios of the future and to inspire people both to think about them and to provoke different reactions to each scenario (Bell 1996a, 316).[9] I take the same approach in this book, where I want to outline some possible stories and inspire readers to think about and further discuss possible future scenarios, assuming they will be determined by space exploration and biomedical technologies.

It is worth emphasizing that reference to the future is important for ethical thinking in general, and for consequentialism in ethics and bioethics in particular (Bell 1996b, 95–96). Hence, space bioethics, as bioethics practiced within futures studies, is consequentialist because it is oriented toward moral evaluation of events (scenarios, predictions) lying in the future. It is worth quoting Bell at this point, who says that "for futurists, morality is not merely a matter of intentions. It is much more a matter of actual results or consequences" (Bell 1996b, 103).[10] Intentions are important, but consequences are more important. The deontological intention to protect autonomy is important as long as it does not lead to negative consequences.

The dispute between consequentialism and deontologism is in space bioethics about which of these theories will better protect human life and which will prevent negative consequences to a lesser extent. But one can turn this question around and say that it is also a question of which of these theories, if accepted as public policy, will prevent the coming into existence and the realization of even better consequences that would have occurred had public policy not been implemented based on this alternative moral theory that prevents the implementation of certain scenarios. In the case of the conflict of differing ethical-political scenarios, the conflict in principle is not between opposing values, but between differing interpretations regarding the prediction of future consequences of particular policies.

Bell believes that we do not differ too much with respect to core values. In his view, the shared axiological minimum is the desire to reduce human suffering and protect life. In contrast, we disagree with respect to which ethical scenario will better protect these values (Bell 1996b, 102–103). Therefore, I propose an issue-driven rather than theory-driven approach to bioethics. I prefer a consequentialist perspective and suggest that consequentialism is particularly close to space bioethics, but I also point to those moments when a deontic perspective seems more appropriate.

Space bioethics as I discuss it in this book is necessarily part of futures studies even if we focus on considering selected biomedical problems in space. Space bioethics itself, on the other hand, does not necessarily fit into futures studies at all if it is understood as considering clinical protocols for currently ongoing research on the International Space Station and addresses issues of autonomy, informed consent, and privacy of astronaut medical data. Viewed in the light of futures studies, this book should be seen as a kind of thought experiment, a form of considering what we would do, what moral decisions we would make, in a situation where a given scenario, even if unrealistic in today's world, becomes real in the future.

If we adopt the mindset inherent in futures studies and realize that events far in the future have their origins and at least partial causes in events and decisions lying in the past, we will understand that neither the future of humans in space nor the large-scale application of human enhancement through biomedical means is the prospect we may already be heading toward. Depending on our preferences, we can accelerate the trajectory of events that lead to these futures, or try to prevent them.

Notes

Chapter 1

1. On the specifics of futures studies and how this book fits into the futures studies framework, see Appendix 1.
2. This assumption is not so obvious, as evidenced by the account of astronaut Scott Kelly, who, when asked if he would be interested in flying to Mars, replied: "I would go, as long as it was a round trip. Having lived on the space station for a year, I would not want to live for the rest of my life in some habitat on Mars" (An Extraordinary Astronaut 2020).
3. See, for example, (Schwartz and Milligan 2016), (Green 2021), where space bioethics is almost never discussed.
4. I treat ethical astrobiological works as belonging to environmental ethics. Only in rare exemptions, other issues are considered which go beyond environmental ethics, see (Smith and Mariscal 2020), (Chon Torres et al. 2021).
5. See the importance of integrity of astronomical objects (Milligan 2015a).
6. See my many articles in collaboration with colleagues (a continuously updated list of publications is available in Szocik 2021e), including a collective book edited by me (Szocik 2020b); see also (Abylkasymova 2021).
7. On what bioethics is, see, among others, (Arras 2020).
8. It is worth noting recent proposals for the widespread use of both astrobioethics (Chon-Torres 2021a) and astroethics (Peters 2021). See also (Chon Torres et al. 2021).
9. I refer readers interested in this knowledge to the professional literature; see, for example (Kieffer et al. 1992), (Carr 1996), 2006), (Kargel 2004), (Forget et al. 2006), (Chapman 2007), (Bell 2008).

Chapter 2

1. The degree of danger of a human being in the space environment is a result not only of currently possessed protective measures, but also of decisions about the goals of further space missions. The size of space means that always some type of mission, due to its duration and distance, will pose a challenge justifying and requiring human enhancement even if the current goal at a given stage of space exploration advancement is achieved without the use of human enhancements.
2. In many of my articles, my colleagues and I have reviewed in detail the specific features of the space environment that make it unsafe for humans, and we have highlighted the differences with Earth. Also the issue of the specific effects of particular environmental factors in space, especially cosmic radiation on human health, has been discussed in detail with colleagues in my following texts, among others (Szocik 2019c, 2019d), (Szocik, Abood, et al. 2018), (Szocik, Elias Marques, et al. 2018), (Szocik, Campa, et al. 2019), (Szocik, Abood, et al. 2020), (Szocik, Wójtowicz, et al. 2020), (Szocik, Shelhamer, et al. 2021), (Szocik, Rappaport, et al. 2021), (Szocik and Braddock 2019). See also (Ball and

Evans 2001, 64–68), (National Council on Radiation Protection and Measurements 2011), (Afshinnekoo et al. 2020), (Wanjek 2020), (Cucinotta and Cacao 2020), (Cucinotta et al. 2021), (Green 2021, 51–56), (Young and Sutton 2021).

3. See my texts, where my colleagues and I, after reviewing existing countermeasures, pointed out their inadequacy for long-term missions and suggested the concept of human enhancement as a hypothetical alternative (Szocik 2015, 2019a, 2020a, 2020b, 2020c, 2020d, 2021a, 2021e, 2023), (Abylkasymova and Szocik 2019), (Szocik and Braddock 2019), (Szocik and Tachibana 2019), (Szocik and Wójtowicz 2019), (Szocik, Norman, et al. 2020), (Szocik, Wójtowicz, et al. 2020), (Szocik, Shelhamer, et al. 2021), (Szocik, Rappaport, et al. 2021), (Mazur et al. 2020).

4. However, some studies show that blocking p53 can in some contexts promotes cancer risk (Lee et al. 2015). This is also a good illustration of the constant danger in the form of side effects of gene editing, many of which we cannot predict today. It is to be hoped that, as our knowledge advances, we will strive to achieve a high degree of certainty regarding the safety of biomedical procedures.

5. See an example of a conservative approach to human enhancement for terrestrial applications (Kamm 2009).

6. Perhaps we should, when considering all possible future scenarios, be open to all options for modifying humans. There would be differences between missions for science, space race, or survival of species that would lead to differences in approaches and ethical considerations. We cannot rule out a scenario in which even the most radical types of biomodification will be applied to even the most trivial missions, if return on investment and cost/benefit analysis determine the feasibility of such an approach.

7. These authors say nothing about space bioethics.

8. Beyond any doubt, what Chadwick and Schüklenk propose is more akin to the classic top-down method, in which we apply principles and rules to individual cases. Nonetheless, what their proposal shares with my approach is methodological openness and uncertainty about what solutions work best under what circumstances. I myself allow, and sometimes propose, a top-down model when I accept as prima facie obligations certain values and principles such as autonomy, respect, and well-being (i.e., a mixed deontological-consequentialist model), but I do not prejudge that any of these principles should be treated in an absolute way.

9. It is also worth adding that when considering bioethical thought experiments, I am simultaneously assuming the possibility of particular future scenarios. We may see some of them as more, and others as less likely. However, this does not change the fact that from today's perspective and knowledge, they are neither true nor false, they are simply possible. See Jan Łukasiewicz's proposal to assign to sentences about the future a third logical value, possibility (1/2), separate from the values of truth (1) and falsity (0) (Łukasiewicz 1961, 153).

10. However, this does not mean that I do not refer to any principles or rules, on the contrary, it is hard to imagine a situation in which we do not consider any principles and rules. Being issue-driven only means that I do not absolutize any principle, which does not change the fact that, paradoxically, many of my considerations may be guided by a particular principle, at least up to a certain point.

11. The method of reflective equilibrium and the search for actual obligations in a given situation are often used by me when analyzing particular biomedical scenarios in space. See more about applying reflective equilibrium, weighing methods and balancing principles

and rules, and the tension between prima facie and actual obligations (Beauchamp and Childress 2013, 15–16, 19–24, 404–410).

12. Rawls's famous concept of veil of ignorance may be a useful tool here (Rawls 1999, 118–123).

13. I think my understanding of space bioethics as issue-driven and case-driven may overlap to some extent with a weak version of rule-utilitarianism. This approach sees moral rules as prima facie obligations, but which can be overridden in certain situations. As I indicate in Appendix 1, futures studies are necessarily a type of consequentialist ethical reflection. In a sense, then, one might say that the methodological profile of space bioethics is shaped by its nature as rooted in futures studies. I also point out in the conclusion of chapter 6 that space bioethics is often consequentialist. It is important to note, however, that this consequentialism, or weak version of rule-utilitarianism, is not a theory-driven approach, but a method of analyzing particular cases and situations and deciding what principles, rules, and obligations should be normative. It is not, therefore, a weak version of rule-utilitarianism in the classical, literal sense, because for utilitarianism, whether act- or rule-utilitarianism, there is always a guiding principle of utility at the starting point, which at most can be mediated by moral rules, which the weak form of rule utilitarianism assumes.

14. As I emphasize in Appendix 1, the precautionary approach typical of bioethics is complemented in the case of space bioethics by a more open-minded and progressive thinking inherent in at least some modes of futures studies.

15. As I mentioned earlier, the nature of space missions is that we can always—because of the vastness of space and the myriad of potential destinations—want to travel farther and farther, beyond the limits of what is currently possible, and what we have already achieved with the countermeasures we currently have.

16. This is a pretty strong claim, which may imply that there are different modes of ethical thinking appropriate to different practices.

17. However not Rawls himself.

18. Pullman talks about respect for human dignity. I am not a proponent of the concept of human dignity, especially in bioethical discussions. It is so vague that it often wrongly leads to stagnant discussions of human enhancement. On the other hand, I consider the principle of respect itself to be substantive, understood as respect for person and respect for rights-holders.

19. Henry T. Greely interestingly describes the negative adventures caused by He Jiankui's experiment for science itself, in the form of a warning for the future, showing what science should not do (Greely 2021).

20. The concept of gaming narratives (Pavarini et al. 2021) seems to have some potential for application to at least some considerations within space bioethics. As gaming narratives refer to realities that no human has yet experienced, are in fact the most perfect, not to say the only reliable and "empirical" tool for bioethics research in future human space missions.

21. See very constructive advice on how to pursue a single preferred ethical theory while being aware of the perhaps inevitable trade-offs (DeGrazia and Millum 2021, 15–20 ff).

22. This is why utilitarianism, and especially the weak version of rule utilitarianism, is often used in public policy, as during the covid-19 pandemic, where as a moral rule, we accept freedom and autonomy but agree to limit them in certain circumstances where leaving these principles intact would do more harm than good.

23. Nevertheless, my approach to bioethics, philosophy, and ethics is pragmatic. It is definitely an applied philosophy, primarily because of the high risk of space missions, as well as the amount of investment and effort put into space exploration (see Rappaport and Szocik 2021).

24. Very few authors have considered within space bioethics the issues discussed in this book, namely human enhancement, especially through GGE for future deep-space missions. Among these exceptions is Christopher E. Mason, who, however, as an eminent geneticist, focuses on the genetic aspect, basically showing the very benefits of radical genetic modification in space (Mason 2021). I refer to Mason's concepts frequently in my book. Brian Patrick Green, on the other hand, mentions in the margins of considering health challenges in space the possibility of genetic modification of astronauts, as well as in the context of ethical challenges such as intergenerational consent or justice in the journey to the planets and stars, bringing into the discussion the very distant prospect of interstellar travel and the idea of the embryonic astronaut, among others (Green 2021, 63, 195–215). Space bioethics understood in this way, however, is not the center of Green's interest, but only one of a dozen chapters in his *Space Ethics*. Therefore, it can be said that the issue of space bioethics understood in this way has not been generally considered in previous publications except for my articles, in which bioethical issues were discussed, if at all, only in passing. On the other hand, the issue of human enhancement in the context of space missions, specifically space colonization, has been mentioned by some philosophers, among others (Bostrom and Savulescu 2009, 19), although it is difficult to say whether they had in mind its application to space colonization or rather how the issue of human enhancement might interact with other global challenges, among which they mentioned space colonization.

25. The principle of respect for rights-holders overlaps not only with the principle of autonomy and freedom, but also with the principle of justice in the Rawlsian understanding of individual freedom as absolutely guaranteed by justice (Rawls 1999, 3).

26. Space bioethics, on the other hand, especially because of its orientation toward the future, is particularly interested in analyzing good consequences and, therefore, what serves well-being. However, not every action that serves well-being, especially when understood at the population level (such as the slogan of survival or saving the human species through space colonization), is in accord with considered moral judgment. This is why the dual value theory is a good tool for balancing the search for well-being, so that it is not a quest to increase well-being at all costs and by all means, but takes into account the balancing principle of respect for rights-holders.

27. See DeGrazia and Millum's observation that, in principle, only rape, enslavement, and torture can be considered absolute moral constraints, in the sense that it is difficult for us to imagine any ethical scenario in which the benefits of performing any of these acts would outweigh the losses and harms (DeGrazia and Millum 2021, 54). I can't think of any scenario where committing rape on another person or making them a slave was actually going to have any good consequences that would justify that rape or enslavement. By contrast, I am skeptical about an absolute prohibition of torture in all situations. One can imagine situations in which torture might be necessary to obtain information on which the survival of others (well-being) might depend.

28. As I point out in Appendix 1, the future may include phenomena that are qualitatively new, unlike and irreducible to any past events. It may also require the application of new methods of problem solving.

Chapter 3

1. See Rachell Powell's interesting piece on the weakening of the gene pool by medicine and the need for human enhancement to fix it (Powell 2015).

Chapter 4

1. The following section is an expanded version of my paper prepared by invitation from *Science* (Szocik 2021a).
2. By this I mean those variants of utilitarianism that can lead to consequences more or less resembling repugnant conclusion (Parfit 1984). Another example is situations in which the concept of sacrificing an individual(s) for the good of the whole community would arise.
3. See similar arguments and conclusions that Koji Tachibana draws from virtue ethics (Tachibana 2020).

Chapter 5

1. On the potential development of a future space habitat into a totalitarian regime, challenges to our fundamental rights, and freedoms, see, among others: (Cockell 2015a, 2015b, 2016a), (Schwartz and Milligan 2016).
2. This division is consistent with my methodological approach, which is bioethics and futures studies, rather than space policy understood as belonging to political science. If we were to adopt the latter perspective, we should probably say that these reasons I list as rationales in favor of space exploration and exploitation, beyond the scientific, such as resources being depleted on Earth (for mining in space) and on benefit/costs of going there, are not really supportable in today's world. Looking historically and politically, we go to space for two basic reasons: finding out more about scientific facts we do not understand—the human curiosity reason—and for reasons of national security and defense. Everything else has no proven justification, but can only be shown to have been good investments after the fact. Conversely, does this realist approach take sufficient account of the power of capitalism, or does it still harbor Cold War thinking, where the only relevant actor was the state? According to some, the main driver of expansion in space is capitalism; moreover, it is capitalism that even underlies the idea of space colonization, as can be seen in the statements of the owners of private companies engaged in space exploration (Gunderson et al. 2021). Martin J. Rees points out that the best inspiration for future space missions will come from private business motivated by the chance for financial gain or inspiration will come from someone's obsessive idea, but certainly not from programs implemented in the name of so-called national or governmental strategies, as was the case during the Cold War (Rees 2003, 176).
3. Nor do I believe, as Robert Zubrin, for example, does, that Mars is our destiny and goal (Zubrin 1996, 297). Humanity may as well never pursue long-term space missions and never go beyond what it is currently pursuing, which is Earth orbit missions. Issues such as the rationale for space missions in light of selected legal, political, or ethical issues were analyzed with coauthors in the following texts, among others: (Szocik 2019c, 2019d), (Szocik et al. 2016, 2017), (Szocik and Tkacz 2018), (Abylkasymova and Szocik 2019), (Mazur et al. 2020), (Szocik, Abood, et al. 2020), (Wójtowicz and Szocik 2021).

4. With any type of mission, regardless of its importance and motivation, as well as humanity's technological preparedness, risk estimation and decision-making under risk will be an important factor. See more about risk and cost-benefit analysis for space missions, as well as difficulties and biases in getting people to correctly assess risk in (Green 2021, chapter 3, chapter 7). On the benefits of long-term space exploration and the technical feasibility of doing so, see (Impey 2015).

5. On the importance of scientific exploration of Mars, see, among others, (National Research Council 2005), (Taylor 2010), (Weintraub 2018).

6. In some basic, operational sense, any robot has the advantage over humans (not just in space, but on Earth) that it can operate in the "three Ds" environment: dull, dirty, and dangerous. Keith Abney adds a fourth D, being dispassionate (Abney 2017, 354).

7. See also my publications where I consider the practical contexts and ethical challenges and implications of choosing either human or uncrewed missions: (Szocik 2019c, 2019d, 2020), (Szocik and Tachibana 2019). The controversy over human or uncrewed missions is a popular topic in many publications and mainly addresses issues such as efficiency, safety, and cost; see, among others, (Coates 1999), (Marshall 2001), (Shelhamer 2017).

8. Abney suggests that sending humans instead of robots on all types of missions, including scientific missions, except for colonization missions may be immoral (Abney 2017, 366).

9. According to some, the discovery of life beyond Earth would be comparable in stature to such cultural and ideological revolutions as the Copernican and Darwinian revolutions (Smith and Mariscal 2020).

10. Both the discovery of traces of life on Mars and the possibility of ruling out any life on Mars will lead to important conclusions about the origin and evolution of life (Walter 1999, 152–154). The scientific significance of these discoveries warrants special concern to avoid forward and backward contamination (Cabrol 2021). See also, other astrobiology research potentially relevant to humanity, not to mention its purely scientific value: (Osinski et al. 2020), (Schulze-Makuch et al. 2020), (Snyder-Beattie et al. 2021), (Limaye et al. 2021).

11. See also the benefits of space missions in (Green 2021, 4).

12. However, it is worth remembering the persuasive arguments of the proponents of commercial exploitation of space, who see in the private space projects currently underway, among other things, a chance to lower the cost of space travel and make it more popular in the future (Crawford 2021).

13. As presented in chapter 1, environmental ethics is arguably today's most important, broadest, and most popular branch of space philosophy and space ethics. See also many publications on space policy, space law, and sustainability in space, which tend to be less critical but highlight the existing risks in immoderate, unregulated space exploitation: (Hertzfeld 2009, 2021), (Hertzfeld and Pace 2013), (Hertzfeld et al. 2016), (Sadeh 2013), (Dallas et al. 2021), (Krag 2021), (Palmroth et al. 2021).

14. But we may never have to consider this scenario, for, as my colleagues and I have suggested in one paper, the exploitation of space may always be accomplished in an uncrewed manner (Campa et al. 2019).

15. There is no doubt that informed consent plays an important role here, for both commercial and scientific missions. It is even possible that people will waive their rights to avoid severe modification.

16. See a critique of thinking about humanity's future in space, and especially thinking about space colonization, from a capitalist profit-seeking perspective. The authors talk about the

"billionaire space race" and the drive for the "militarization of space and privatization of space travel": (Gunderson et al. 2021).

17. On the challenges and expected benefits of space tourism, see, among others, (Van Pelt 2005), (Gibson 2012), (Toivonen 2021).

18. Since the existing space tourists are elderly, it can be presumed that their health and fitness are even lower than the population average, as well as exposing them more to the hazards of the space environment.

19. However, this is, as I mentioned, a special assumption for the sake of the argument. For in reality, it seems that the very rich, a prime demographic for space in the foreseeable future, will be able to readily get even frontier forms of modification.

20. At this point I should add a skeptical note. I am well aware that the space refuge concept is a long way off and may never be realized. Many experts and scientists, probably rightly so, take a very skeptical approach to the idea that any sort of colonization or community of human beings as we know them will be able to live and survive in outer space. This is what probably will not happen in the life time of any living on Earth right now. There are just too many survival issues ranging from radiation to low gravity for which we have no medical knowledge of how to overcome them. Most medical problems have been dealt with by engineers and not by doctors or medical research. For example, they try to harden the capsule to avoid radiation, not adjust astronaut's bodies directly. Or, they provide exercises and special equipment to overcome the effects of zero gravity rather than affect astronauts' internal organs directly. Finally, we can say, adopting a skeptical, realist, and political optics, that the concept of space refuge should not be treated from today's perspective as an issue that solves any problems in space affairs today. On the other hand, in line with my paradigm presented in chapter 1 and Appendix 1, I see the space refuge concept as an interesting thought experiment, but also as one possible future scenario.

21. When I talk about the concept of space refuge or space settlement, I mean a reasonably realistic concept of settling some part of the human population on some other space object that will be realistic to achieve, such as Mars. However, I am not referring to the concept of so-called interstellar travel or interstellar exploration, see (Crawford 2012, 2014).

22. If we take a very realist perspective, oriented toward the here and now, and dispense with the futures studies perspective discussed in Appendix 1, which I use in the book, we should perhaps say that the prospect of space refuge for more than a tiny fraction of the planet's population is very far off, with no imminent moral quandary. The moral challenge, on the other hand, will be at least the selection of the small fraction of the population that is chosen to move to the space colony. Even if we assume that a maximum of 1 percent of the population can be saved by space colonization, for the population living on Earth in 2050 this would be 100 million with an estimated population of 10 billion (Szocik 2021f, 210).

23. As a default position, I accept the idea that we should be concerned about the survival of the species. By contrast, how and by what means, and whether at all costs, remains unclear. In chapter 8, I discuss this problem from the perspective of quality-of-life ethics and the counterarguments of antinatalism.

24. I have discussed these and other issues related to the ethics of quality of life in space, the rationale for the concept of space refuge, and the concept of global catastrophic risks, among others, in the following texts: (Szocik 2019d), (Szocik, Abood, et al. 2020). See also the extensive literature on global catastrophic and existential risks, especially (Ord 2020), as well

as institutes dedicated to this topic (the Global Catastrophic Risk Institute or the Centre for the Study of Existential Risk at the University of Cambridge).

25. Check out this interesting article by John Gowdy, in which the author predicts that a future climate catastrophe on Earth will lead to the collapse of civilization and force a return to the hunter-gatherer lifestyle: (Gowdy 2020).

26. Another key principle besides well-being, the principle of respect for rights-holders, can both reinforce the principle of well-being and remain in limbo for future generations as long as we do not recognize future people as rights-holders.

27. Space colonization is not only a great challenge because of the technical problems and scale of the undertaking. The specific type of difficulty here is that it is a multigenerational project, requiring a way of organizing and financing that goes beyond the thinking, decision-making, and living of a single generation (Haqq-Misra 2019), (Kurtz 2021).

28. See my discussion in chapter 3.

29. However, I do not underestimate the importance of logistical and political problems. On the contrary, they are no less important than the ethical issues discussed here. It is worth remembering that any concept of true and effective international cooperation at any stage of a permanent human base in space will be highly problematic if viewed from today's perspective. On top of that, there are the supply issues. Today and for the foreseeable future anyone in space is tied to a nation or group of nations on Earth, both legally and for support services, food, equipment, etc. This dependence on the country/countries organizing the mission has to be cleanly broken for any true community of people living in space.

Chapter 6

1. This is the opinion of Rees, among others, who points out that the ocean travelers were at a disadvantage compared to the participants of the future expedition to Mars, they had much less knowledge of what awaited them and where they were going, they had no contact or the message they sent took months to reach the recipient (Rees 2003, 176).

2. As I emphasize in chapter 2, these are important issues, and there are many elements regarding clinical bioethics in the space so understood that need improvement. Nevertheless, the topic of my book and the approach inherent in futures studies keep me focused on hypothetical bioethical challenges centered around human enhancement.

3. This is an important assumption. Space bioethics, even if we assume the existence of significant differences between the environment of space missions and even extreme environments on Earth, does not imply the necessity to search for new principles and rules or even to modify the already known principles and rules. Space bioethics assumes the use of existing principles and rules (e.g., principles such as beneficence and autonomy and rules such as informed consent and confidentiality) in specific conditions. What is new, therefore, is the use of known principles and rules, which in various situations of space missions should perhaps be suspended, perhaps applied, perhaps against our moral intuitions and habits. As I emphasized in chapter 2, I take as my main two values and principles the principle of well-being (beneficence, nonmaleficence, and to some extent the principle of utility) and the principle of respect for rights-holders (correlated with the principles of autonomy and freedom).

4. New technologies force us to ask whether the new moral environment that technologies generate always requires only flexibility in terms of which principles and rules and when

and in what combinations we will apply (which I assume in general, and with respect to space bioethics in particular), or whether it may also require the development of new principles and rules, or possibly their modification or reformulation. I am not ruling out the latter possibility as a hypothetical option, but I think we are probably always dealing with option one of the above. See an interesting concept by David Grant, who argues that neuroscience is forcing us to reformulate our thinking about the privacy rule. It is not so much about suspending it in favor of other principles, but simply modifying the principle itself (Grant 2021).

5. By stipulating that a modification does not necessarily have to be medical to be considered a transhumanist modification, Cabrera unnecessarily weakened her definition. For we can envision a situation in which the one and only human on Earth obtains a degree of a given trait in a way that goes well beyond the species typical limit, for example, through hard training.

6. On whether speciation in a space colony would be possible, see more in chapter 3.

7. If the interference were only to correct DNA in individuals and not to modify gametes, then we would avoid affecting the next generation.

8. I point out here that the application of human enhancements to space missions for at least the first generation of modified astronauts will remain unknown to some extent. One can imagine a situation in which some humans may naturally (without enhancement) be tolerant of much higher levels of cosmic radiation or more resistant to conditions of altered gravity, or isolation and distance from Earth than others.

9. The term "unique property" has a contextual meaning here. Since we all have DNA repair mechanisms, an astronaut can obtain at most quantitatively more efficient repair mechanisms, compared to the unmodified population on Earth.

10. I am not including the evolutionary validity of reproduction, without which natural selection makes no sense (Rothman 2015).

11. Clarification is essential here. I argue that the primary purpose of applying human enhancements to space missions will be to protect the health and lives of astronauts. Only secondarily, enhancements may be directed to increase only performance, without reference to health protection. It does not change the fact, however, that every human enhancement, even health related, at the same time increases the performance as long as a healthy astronaut is more efficient than an ill or permanently weak astronaut, which she could be during the mission without human enhancement.

12. This leaves a great moral dilemma as to whether human enhancement applied for the purpose of gaining a competence advantage over others can be morally acceptable in the name of freedom and the right to development, or whether some form of social equality and justice should take precedence.

13. I am not saying that moral bioenhancement on Earth is not needed, quite the contrary. I am referring to the context of the discussion of moral bioenhancement on Earth, where conventional forms of moral enhancement are pointed to as a counterargument. In contrast to Earth, the cosmos is understood as a new moral ecology for which we have not developed any conventional forms of moral enhancement, hence the justification for moral bioenhancement is automatically stronger.

14. But only with respect to local challenges. At the global level, we still have not solved problems like global warming, helping migrants, or ending armed conflicts. In 2020, there were fifty-six ongoing armed conflicts in the world (Conflict Trends 2021). The only thing that

is new is that today they are not taking place on the territory of Western countries, which nevertheless actively participate in many armed conflicts that have been exported outside their territories.

15. For now, we are talking about the kind of space missions we know from past history, but there is a strong case for applying enhancement even to hypothetical massive space missions in the distant future.

16. This sounds extremely improbable, however, given that (1) today consensus on these issues on Earth within a single country/cultural circle is unattainable, and (2) there is a lack of consensus on much less controversial bioethical issues on space missions already underway, which can only increase with multicultural missions.

17. However, it is worth noting that there is a variation between the various Christian churches about the moral acceptability of abortion.

18. One could say that freedom and autonomy overlap with or are simply rooted in the principle of respect for rights-holders.

19. Simplifying, we can say that it is better to have well-being with limited autonomy and freedom over time than autonomy and freedom leading to limited well-being. A good example of this equation is the restrictions placed on our freedom and autonomy during the Covid-19 pandemic. But this equation also has its dark paternalistic side when various governments can, in the name of well-being, impose unreasonable restrictions on our freedom and autonomy. Despite this paternalistic threat, I can hardly imagine any scenario in space where the principle of autonomy should be given the status of actual obligation over well-being.

20. Green points out that sometimes fully informed consent in space may not be possible due to a lack of full knowledge of all the possible effects of being in space. Thus, we are talking here about incomplete informed consent, or informed consent that is not truly informed consent. Another case, however, is the situation preceding the Space Shuttle Challenger disaster, where a fault related to cold rubber O-rings was known, but despite that, a decision was made to launch the space shuttle. However, the crew did not know about the malfunction, so their informed consent to participate in the mission was not full (it was not informed). Finally, informed consent in space can take the form of proxy consent, where some institution with best-informed status will make decisions on behalf of astronauts (Green 2021, 40, 58, 63).

21. As the authors of the famous "Safe Passage" report argue, this issue should be looked at not only in the context of concerns about the right to participate in the next mission but also in the context of selection criteria that are unclear to astronauts, as well as NASA's people culture, stoicism, and a "can-do" belief reluctant to share information about their health (Ball and Evans 2001, 174–175).

22. The assumption here is that the degree of danger, novelty, and unpredictability of the space environment is so high that the principle of privacy and confidentiality become less important, especially in light of a small research group (Ball and Evans 2001, 179–180).

23. See more about autonomy, informed consent, and duty to obey in military ethics in (Messelken and Winkler 2020).

24. However, it is worth remaining skeptical and cautious about those scenarios that will propose such a model, especially since history cautions against such an approach.

25. The book does not discuss the well-known and well-developed issues of risk calculations, probability of harm, risk assessment, or risk management, or the issues of uncertainty and

the precautionary principle, which appear here in the background, but are only a kind of technicality (Mepham 2008, 327–342). See also the health risk assessment strategies used by NASA, as well as predicting medical events in space using the frequency of medical events on submarines and Antarctic expeditions (Ball and Evans 2001, 80–88).

26. This is so specific to many dual-agency settings where the physician has responsibilities to both the patient and the institution, hence the frequent signing of privacy statements where the parties accept and are aware of these specific circumstances. For future long-term space missions of unknown risk, astronaut safety may require the need to disclose and share medical information for improving countermeasures for other astronauts (Ball and Evans 2001, 175–176, 178).

27. Indeed, according to battlefield bioethics, triage is conducted not on the basis of current medical need, but by the criterion of battlefield combat efficiency, that is, as many soldiers as possible being able to return to the battlefield (Gross 2006).

28. See also similar comments regarding potential differences in prevention, treatment, and research participation: (Ball and Evans 2001, 186).

29. It seems that humanity must accept some irreducible risk of fatal accidents in space if it intends to pursue a space program. But this risk can be compared to the standard risks currently assumed for air travel or automobile travel. Conversely, if the possible risk is estimated to be more than standard, one might consider postponing the mission until effective countermeasures are proposed. Green points out that one of the dilemmas and problems that characterize space missions will be sending some group of people into inferior conditions at least early in the space mission program. The living conditions of this group will probably be worse than they would have been had they remained on Earth. But humanity, in turn, may see this as necessary, a form of risky sacrifice for those first long-term astronauts (Green 2021, 205–206). This situation is akin to the risks associated with the first generation of people subjected to GGE for the first time in the future. Do we have the right to sacrifice these people for the sake of humanity? It seems that the justification for a space mission must be strong and nontrivial, as in the space refuge concept. In contrast, as Green points out, the risk of death and total failure is an inevitable part of space programs (Green 2021, 40). See also the different methods for determining risk acceptability (the good-faith subjective standard, the reasonable-person standard, and the objective standard) and the difficulty of implementing each for long-term space missions (Abney 2017, 362).

30. In the colloquial understanding, when talking about the good of the group, the group is usually seen as an independent entity, the good of which is not reducible to the sum of the good of its members. Utilitarianism on the other hand is usually focused on the aggregated good of the individuals (Bentham 2003, 18). Perhaps we should never talk about something like the good of the group, only the good of individuals. But it is hard to ignore the intuitions of Aristotle, who said: "For even if the good is the same for an individual as for a city, that of the city is obviously a greater and more complete thing to obtain and preserve. For while the good of an individual is a desirable thing, what is good for a people or for cities is a nobler and more godlike thing" (Aristotle 2014, 4).

31. But some specificity seems inevitable here. It is difficult to take as an absolute moral principle the principle of preserving life if we cannot exclude the risk of fatal accidents. By contrast, an absolute moral constraint for me is to violate the principle of respect for rights-holders and create conditions in space in which individual rights and freedoms are permanently violated. Thus, the concept of sacrificing an individual has different meanings, and

perhaps sometimes a certain form of sacrifice may be acceptable, but in turn others should never be allowed.

32. Some of them are incommensurate, but at least we should assume the validity of each principle at the outset and approach them with care.

33. See on autonomy understood as freedom from any influence, which guarantees the ability to act voluntarily as well as to act on the basis of values one shares, (DeGrazia and Millum 2021, 100).

Chapter 7

1. To my knowledge, the only papers discussing this issue are as follows: (Szocik 2019a), (Szocik 2020a), (Szocik and Gouw, 2023).

2. To some extent, such modification in space could be considered therapeutic, at least as long as behavioral health is considered one of the most important medical challenges during long-term missions (see Ball and Evans 2001, 137–170).

3. Perhaps the most famous example of the gap between philosophical considerations and what real and living people want is the repugnant conclusion showing the paradoxical effects of one variant of utilitarianism (Parfit 1984, 388).

4. However, it is worth bearing in mind that this argumentation based on the idea of evolutionary mismatch is criticized (see Segovia-Cuéllar and Del Savio 2021).

5. It is worth remembering that social and cultural evolution are two other important forces shaping human development besides biological evolution, which, as a result of their interaction with each other and their interaction with biological evolution, have led to the evolution of many cultural mechanisms regulating behavior; see (Ross and Richerson 2014), (Richerson et al. 2021).

6. We can imagine that Holocaust-level events will happen cyclically in human history. Let us also assume that we cannot perceive this because we are only at the beginning of the process. If humanity were able to recognize this pattern, the moral case for moral bioenhancement would not only be strong but overwhelming, as long as such enhancement actually worked (thanks to Tony Milligan for this example).

7. As I mentioned in the previous chapter, while the diversity of physiological and immune traits to cosmic radiation and the effects of altered gravity in the human population is inadequate, and justifies applying therapeutic enhancements equally to all, it can be assumed that we have quite a bit of diversity in the human population on Earth regarding morality. Perhaps this diversity is large enough that individuals with certain moral traits from one end of this distribution will be sufficiently immune to negative environmental factors in space that they will not require moral bioenhancement.

8. See also the analysis of the potential difficulties of reconciling utilitarian thinking about moral bioenhancement with common morality, especially the role played by rapid decision-making based on moral intuitions, particularly troublesome for act-utilitarianism (Kudlek 2022).

9. See also Anomaly's interesting discussion of the fact that there are rationales for recognizing that cognitive enhancement itself should be supported by moral enhancement (and not necessarily vice versa). Anomaly also points out that despite the potential risks generated by cognitive enhancement, the fact remains that smarter people build more durable, safer, and fairer societies (Anomaly 2020, 8).

10. Finally, it is worth quoting another valuable comment Tony Milligan shared with me regarding the issue of moral bioenhancement. Milligan points out that worries about moral freedom seem to presuppose that moral biomodification would make us like automata. But perhaps it would be more like a shift in our position on the primate evolutionary tree, moving us a little further away from chimps and a little closer to bonobos. On the one hand, we can say that moral bioenhancement would simply be a narrowing of our horizon of moral and behavioral possibilities, but a narrowing that perhaps we should not regret if we were to become actually closer to bonobos (assuming there are no side effects on the individual and society). But, on the other hand, perhaps it would not even be a narrowing of moral and behavioral possibilities, but just a shift with some behavior patterns taken away, but others added.

Chapter 8

1. See the famous former one-child policy in China and the effect it had on limiting population growth.
2. For more on repugnant conclusion see in (Arrhenius et al. 2017), (Parfit 2017).

Appendix

1. We can hope that GGE never brings negative consequences that, in many cases, humanity simply could not have foreseen. Clinton Andrews gives some examples of discoveries from the recent past that have produced adverse effects, such as chlorofluorocarbon refrigerants (1930), thalidomide sedative (1957), and Facebook newsfeed (2006) (Andrews 2021).
2. See (Bell 1996a, 87–88), where Bell emphasizes that the ethical dimension of futures studies is an integral part of it.
3. It is worth complementing these two values with respect for rights-holders (DeGrazia and Millum 2021, 5).
4. Such a thought experiment is proposed by Wendell Bell, among others, but he does not refer to a future in space (Bell 1996a, 116–117).
5. Among the nine main tasks of futures studies listed by Wendell Bell are the following: exploring possible and probable futures, examining images of the futures, orienting the present, or communicating and advocating particular images of the future (Bell 1996a, 75–97).
6. This was indeed the case, and many scientists doubted the possibility of flying with airplanes or the possibility of space flight shortly before these achievements (Bell 1996a, 79).
7. Some have suggested using the term "forecasting" instead of "predictions" within futures studies.
8. Bell lists nine key assumptions (Bell 1996a, 140–157). I consider the assumptions about the possible singularity of the future, about the most useful knowledge, about the open future, the assumption about "humans make themselves," and the assumption about better futures to be the most important.
9. The scenario is my main methodological tool in terms of a futures studies perspective (see Bell 1996a, 317), independent of the methods specific to bioethics discussed in chapter 2.
10. This consequentialist orientation is also evident in the definition of futures studies proposed by Jennifer M. Gidley: "Futures Studies is the art and science of taking responsibility for the long-term consequences of our decisions and our actions today" (Gidley 2017, 136).

References

Abney, K. 2017. Robots and space ethics. In P. Lin, K. Abney, and R. Jenkins (Eds.), *Robot ethics 2.0: From autonomous cars to artificial intelligence* (pp. 354–368). Oxford University Press.

Abney, K., and P. Lin. 2015. Enhancing astronauts: The ethical, legal and social implications. In J. Galliott (Ed.), *Commercial space exploration: Ethics, policy and governance* (pp. 245–257). Ashgate.

Abylkasymova, R. 2021. Book review. *Human enhancements for space missions, lunar, martian, and future missions to the outer planets*, Konrad Szocik (Ed.), Springer International Publishing (2020). *Space Policy* 57: 101443.

Abylkasymova, R., and K. Szocik. 2019. Ethical issues in a human mission to Mars. *Space Research Today* 206: 44–50.

Afshinnekoo, E., et al. 2020. Fundamental biological features of spaceflight: Advancing the field to enable deep-space exploration. *Cell* 183 (5): 1162–1184.

Agar, N. 2010. *Humanity's end: Why we should reject radical enhancement.* MIT Press.

Agar, N. 2014a. There is a legitimate place for human genetic enhancement. In A. L. Caplan and R. Arp (Eds.), *Contemporary debates in bioethics* (pp. 343–352). John Wiley & Sons.

Agar, N. 2014b. *Truly human enhancement: A philosophical defense of limits.* MIT Press.

Agar, N., and F. Marshall. 2015. Human enhancement. In J. D. Arras, E. Fenton, and R. Kukla (Eds.), *The Routledge companion to bioethics* (pp. 531–542). Routledge.

Allyse, M., Y. Bombard, R. Isasi, et al. 2019. What do we do now? Responding to claims of germline gene editing in humans. *Genetics in Medicine* 21: 2181–2183.

Alonso, M., J. Anomaly, and J. Savulescu. 2020. Gene editing: Medicine or enhancement? *Ramon Llull Journal of Applied Ethics* 11: 259–276.

Alonso, M., and J. Savulescu. 2021. He Jiankui's gene-editing experiment and the non-identity problem. *Bioethics* 35: 563–573.

An extraordinary astronaut. 2020. *Cell* 183 (6): 1457–1461.

Andrews, C. 2021. *Better anticipating unintended consequences.* Presentation given at an online meeting of the Yale Technology and Ethics group, November 11, 2021.

Anomaly, J. 2018. Defending eugenics: From cryptic choice to conscious selection. *Monash Bioethics Review* 35: 24–35

Anomaly, J. 2020. *Creating future people: The ethics of genetic enhancement.* Routledge, Taylor & Francis Group.

Anomaly, J., and G. Jones. 2020. Cognitive enhancement and network effects: How individual prosperity depends on group traits. *Philosophia* 48: 1753–1768.

Aristotle. 2014. *Nicomachean ethics.* Translated and edited by Roger Crisp. Cambridge University Press.

Arnould, J. 2011. *Icarus' second chance: The basis and perspectives of space ethics.* Springer.

Arnould, J. 2017. *Impossible horizon: The essence of space exploration.* ATF Press.

Arras, J. 2020. Theory and bioethics. In Edward N. Zalta (Ed.), *Stanford encyclopedia of philosophy*, Fall 2020 Edition. https://plato.stanford.edu/archives/fall2020/entries/theory-bioethics/.

Arras, J. D. 2009. A case approach. In H. Kuhse and P. Singer (Eds.), *A companion to bioethics* (2nd ed., pp. 117–125). Wiley-Blackwell.

Arras, J. D. 2017. *Methods in bioethics: The way we reason now.* Edited by James Childress and Matthew Adams. New York: Oxford University Press.

Arrhenius, G., J. Ryberg, and T. Tännsjö. 2017. The repugnant conclusion. In Edward N. Zalta (Ed.), *The Stanford encyclopedia of philosophy,* Spring 2017 Edition. https://plato.stanford.edu/archives/spr2017/entries/repugnant-conclusion/.

Asimov, I. 1956. *I, robot.* New American Library.

Balistreri, M., and S. L. Hansen. 2019. Moral and fictional discourses on assisted reproductive technologies: Current responses, future scenarios. *Nanoethics* 13: 199–207.

Ball, J. R., and C. H. Evans Jr. (Eds.). 2001. *Safe passage: Astronaut care for exploration missions.* Committee on Creating a Vision for Space Medicine during Travel beyond Earth Orbit. Board on Health Sciences Policy. Institute of Medicine. National Academy Press.

Barclay, L. 2016. A natural alliance against a common foe? Opponents of enhancement and the social model of disability. In S. Clarke, J. Savulescu, C. A. J. Coady, A. Giubilini, and S. Sanyal (Eds.), *The ethics of human enhancement: Understanding the debate* (pp. 75–86). Oxford University Press.

Barfield, W. 2019. The process of evolution, human enhancement technology, and cyborgs. *Philosophies* 4 (1): 10: 1–14. https://doi.org/10.3390/philosophies4010010.

Barfield, W., and A. Williams. 2017. Cyborgs and enhancement technology. *Philosophies* 2: 4. https://doi.org/10.3390/philosophies2010004.

Barfield, B., and A. Williams. 2017. Law, cyborgs, and technologically enhanced brains. *Philosophies* 2 (1): 6: 1–17. https://doi.org/10.3390/philosophies2010006.

Battisti, D. 2021. Affecting future individuals: Why and when germline genome editing entails a greater moral obligation towards progeny. *Bioethics* 35: 487–495. https://doi.org/10.1111/bioe.12871.

Baum, S. D. 2016. The ethics of outer space: A consequentialist perspective. In J. Schwartz and T. Milligan (Eds.), *The ethics of space exploration* (pp. 109–123). Springer.

Baxter, J. 2021. When is it safe to edit the human germline? *Science and Engineering Ethics* 27: 43: 1–21. https://doi.org/10.1007/s11948-021-00320-x.

Baylis, F. 2017. Human germline genome editing and broad societal consensus. *Nature Human Behaviour* 1: 103. https://doi.org/10.1038/s41562-017-0103.

Baylis, F. 2018. The potential harms of human gene editing using CRISPR-Cas9. *Clinical Chemistry* 64 (3): 489–491.

Baylis, F. 2019. *Altered inheritance: CRISPR and the ethics of human genome editing.* Harvard University Press.

Beauchamp, T. L., and J. F. Childress (Eds.). 2013. Principles of biomedical ethics (7th ed.). Oxford University Press.

Beech, M., J. Seckbach, and R. Gordon. 2021. *Terraforming Mars.* Astrobiology Perspectives on Life of the Universe. Scrivener Publishing—Wiley.

Bell, J. (Ed.). 2008. *The Martian surface: Composition, mineralogy, and physical properties.* Cambridge University Press.

Bell, W. 1996a. *Foundations of futures studies: Human science for a new era. V. 1. History, purposes, and knowledge.* Transaction Publishers.

Bell, W. 1996b. *Foundations of futures studies: Human science for a new era. V. 2. Values, objectivity, and the good society.* Transaction Publishers.

Belshaw, C. 2021. *The value and meaning of life.* Routledge, Taylor & Francis Group.

Benatar, D. 2006. *Better never to have been: The harm of coming into existence.* Clarendon Press.

Benatar, D. 2015. Anti-natalism, In D. Benatar and D. Wasserman (Eds.), *Debating procreation: Is it wrong to reproduce?* (pp. 9–132). Oxford University Press

Benatar, D., and D. Wasserman (Eds.). 2015. *Debating procreation: Is it wrong to reproduce?* Oxford University Press.

Bentham, J. 2003. An introduction to the principles of morals and legislation. In J. Stuart Mill, *"Utilitarianism" and "On liberty" including Mill's "Essay on Bentham," and selections from the writings of Jeremy Bentham and John Austin*, edited with an introduction by M. Warnock (pp. 17–51). Blackwell.

Berry, R. M. 2007. *The ethics of genetic engineering*. Routledge, Taylor & Francis Group.

Billings, L. 2017. Should humans colonize other planets? No. *Theology and Science* 15 (3): 321–332.

Billings, L. 2019. Colonizing other planets is a bad idea. *Futures* 110: 44–46.

Blackford, R. 2014. *Humanity enhanced: Genetic choice and the challenge for liberal democracies*. MIT Press.

Blackshaw, B., and D. Rodger. 2021. If fetuses are persons, abortion is a public health crisis. *Bioethics* 35: 465–472. https://doi.org/10.1111/bioe.12874.

Blodgett-Ford, S. J. 2021. Human enhancements and voting: Towards a declaration of rights and responsibilities of beings. *Philosophies* 6 (1): 5: 1–43. https://doi.org/10.3390/philosophies6010005.

Boggio, A., and R. Yotova. 2021. Gene editing of human embryos is not contrary to human rights law: A reply to Drabiak. *Bioethics* 35: 956–963, https://doi.org/10.1111/bioe.12945.

Bognar, G. 2019. Overpopulation and procreative liberty. *Ethics, Policy, and Environment* 22 (3): 319–330.

Bostrom, N., and R. Roache. 2008. Ethical issues in human enhancement. In J. Ryberg, T. Petersen, and C. Wolf (Eds.), *New waves in applied ethics* (pp. 120–152). Palgrave Macmillan.

Bostrom, N., and A. Sandberg. 2009. The wisdom of nature: An evolutionary heuristic for human enhancement. In J. Savulescu and N. Bostrom (Eds.), *Human enhancement* (pp. 375–416). Oxford University Press.

Bostrom, N., and J. Savulescu. 2009. Introduction: Human enhancement ethics; The state of the debate. In J. Savulescu and N. Bostrom (Eds.), *Human enhancement* (pp. 1–22). Oxford University Press.

Boulter, M. 2002. *Extinction: Evolution and the end of man*. Columbia University Press.

Boylan, M., and K. E. Brown. 2001. *Genetic engineering: Science and ethics of the new frontier*. Prentice Hall.

Brock, D. W. 2009. Is selection of children wrong? In J. Savulescu and N. Bostrom (Eds.), *Human enhancement* (pp. 251–276). Oxford University Press.

Brown, F. L., and L. A. Keefer. 2020. Anti-natalism from an evolutionary psychological perspective. *Evolutionary Psychological Science* 6 (3): 283–291.

Buchanan, A., and R. Powell. 2018. *The evolution of moral progress: A biocultural theory*. Oxford University Press.

Buchanan, A. E. 2011a. *Better than human: The promise and perils of enhancing ourselves*. Oxford University Press.

Buchanan, A. E. 2011b. *Beyond humanity? The ethics of biomedical enhancement*. Oxford University Press.

Cabrera, L. Y. 2015. *Rethinking human enhancement: Social enhancement and emergent technologies*. Palgrave Macmillan.

Cabrera, L. Y. 2017. Reframing human enhancement: A population health perspective. *Front Sociol* 2:4.

Cabrol, N. A. 2021. Tracing a modern biosphere on Mars. *Nature Astronomy* 5: 210–212.

Callahan, D. 2003. *What price better health? Hazards of the research imperative*. University of California Press.

Callahan, D. 2004. Bioethics. In S. Garrard (Ed.), *The encyclopedia of bioethics* (3rd ed.). Macmillan.

Campa, R., K. Szocik, and M. Braddock. 2019. Why space colonization will be fully automated. *Technological Forecasting and Social Change* 143: 162–171.

Caplan, A. 2004. Arthur Caplan's viewpoint: Nobody is perfect—but why not try to be better? In A. Caplan and C. Elliott (Eds.), Is It Ethical to Use Enhancement Technologies to Make Us Better Than Well? *PLoS Medicine* 1 (3) e69 e52: 172–175.

Caplan, A. 2019. Getting serious about the challenge of regulating germline gene therapy. *PLoS Biology* 17 (4): e3000223. https://doi.org/10.1371/journal.pbio.3000223.

Carman, M. 2021. The limits of direct modulation of emotion for moral enhancement. *Bioethics* 35: 192–198.

Carr, M. H. 1996. *Water on Mars*. Oxford University Press.

Carr, M. H. 2006. *The surface of Mars*. Cambridge University Press.

Chadwick, R. F., and U. Schüklenk. 2021. *This is bioethics: An introduction*. John Wiley & Sons.

Chan, S. 2019. Commentary on "Moral reasons to edit the human genome": This is not the moral imperative we are looking for. *Journal of Medical Ethics* 45: 528–529.

Chance, B. A. 2021. Kant and the enhancement debate: Imperfect duties and perfecting ourselves. *Bioethics* 35: 801–811, https://doi.org/10.1111/bioe.12906.

Chang, P. L (Ed.). 1995. *Somatic gene therapy*. CRC Press.

Chapman, M. G. (Ed.). 2007. *The geology of Mars: Evidence from Earth-based analogs*. Cambridge University Press.

Childress, J. F. 2009. A principle-based approach. In H. Kuhse and P. Singer (Eds.), *A companion to bioethics* (2nd ed., pp. 67–76). Wiley-Blackwell.

Chon-Torres, O. A. 2018. Astrobioethics. *International Journal of Astrobiology* 17: 51–56.

Chon-Torres, O. A. 2020. Moral challenges of going to Mars under the presence of non-intelligent life scenario. *International Journal of Astrobiology* 19 (1): 49–52.

Chon-Torres, O. A. 2021a. Astrobioethics: Epistemological, astrotheological, and interplanetary issues. In O. A. Chon-Torres, T. Peters, J. Seckbach, and R. Gordon (Eds.), *Astrobiology: Science, ethics, and public policy* (pp. 1–15). Wiley-Scrivener Publishing.

Chon-Torres, O. A. 2021b. Disciplinary nature of astrobiology and astrobioethic's epistemic foundations. *International Journal of Astrobiology* 20: 186–193.

Chon-Torres, O. A. 2021c. Mars: A free planet? *International Journal of Astrobiology* 20 (4): 294–299, https://doi.org/10.1017/S1473550421000161.

Chon-Torres, O. A., T. Peters, J. Seckbach, and R. Gordon (Eds.). 2021. *Astrobiology: Science, ethics, and public policy*. Wiley-Scrivener Publishing.

Clark, D. P., and N. J. Pazdernik. 2009. *Biotechnology: Applying the genetic revolution*. Academic Press. Elsevier.

CNN. 2021. *Prince William slams space tourism and says billionaires should focus on saving Earth*. By Hannah Ryan, CNN Business; updated 1:48 PM ET, Thu October 14, 2021. https://www.cnn.com/2021/10/14/business/prince-william-space-tourism-intl-scli/index.html.

Coady, C. A. J. 2009. Playing God. In J. Savulescu and N. Bostrom (Eds.), *Human enhancement* (pp. 155–180). Oxford University Press.

Coates, A. J. 1999. Limited by cost: The case against humans in the scientific exploration of space. *Earth, Moon, and Planets* 87: 213–219.

Cockell, C. S. (Ed.). 2015a. *Human governance beyond Earth: Implications for freedom*. Springer.

Cockell, C. S. (Ed.). 2015b. *The meaning of liberty beyond Earth*. Springer.

Cockell, C. S. (Ed.). 2016a. *Dissent, revolution and liberty beyond Earth*. Springer.

Cockell, C. S. 2016b. The ethical status of microbial life on earth and elsewhere: In defence of intrinsic value. In J. S. J. Schwartz and T. Milligan (Eds.), *The ethics of space exploration* (pp. 167–179). Springer.

Cohen, C. B. 1996. "Give me children or I shall die!" New reproductive technologies and harm to children. *Hastings Center Report* 26 (2): 19–27.

Conflict trends. 2021. *Trends in armed conflict, 1946–2020*. Peace Research Institute Oslo (PRIO).

Congressional Research Service. 2018. *Advanced gene editing: CRISPR-Cas9*. Congressional Research Service. https://crsreports.congress.gov, R44824.

Crawford, I. A. 2012. Stapledon's interplanetary man: A commonwealth of worlds and the ultimate purpose of space colonisation. *Journal of the British Interplanetary Society* 65: 13–19.

Crawford, I. A. 2014. Avoiding intellectual stagnation: The starship as an expander of minds. *Journal of the British Interplanetary Society* 67: 253–257.

Crawford, I. A. 2021. *Private space expansionism: Potential scientific and societal benefits and implications for governance*. World Orders Forum, World Government Research Network.

Crusan, J. C., et al. 2018. Deep space gateway concept: Extending human presence into cislunar space. 2018 IEEE Aerospace Conference, IEEE, pp. 1–10, 10.1109/AERO.2018.8396541.

Cucinotta, F. A., and E. Cacao. 2020. Predictions of cognitive detriments from galactic cosmic ray exposures to astronauts on exploration missions. *Life Sciences in Space Research* 25: 129–135.

Cucinotta, F. A., W. Schimmerling, E. A. Blakely, and T. K. Hei. 2021. A proposed change to astronaut exposures limits is a giant leap backwards for radiation protection. *Life Sciences in Space Research* 31: 59–70.

Dale, J. W., and M. von Schantz. 2007. *From genes to genomes: Concepts and applications of DNA technology* (2nd ed.). John Wiley & Sons, Ltd.

Dallas, J. A., S. Raval, S. Saydam, and A. G. Dempster. 2021. An environmental impact assessment framework for space resource extraction. *Space Policy* 57: 101441.

Daniels, N. 2009. Can anyone really be talking about ethically modifying human nature? In J. Savulescu and N. Bostrom (Eds.), *Human enhancement* (pp. 25–42). Oxford University Press.

DeGrazia, D. 2012. *Creation ethics: Reproduction, genetics, and quality of life*. Oxford University Press.

DeGrazia, D., and J. Millum. 2021. *A theory of bioethics*. Cambridge University Press.

de Lazari-Radek, K., and P. Singer. 2017. *Utilitarianism: A very short introduction*. Oxford: Oxford University Press.

Des Marais, D. J., B. M. Jakosky, and B. M. Hynek. 2008. Astrobiological implications of Mars's surface composition and properties. In J. Bell (Ed.). *The Martian surface: Composition, mineralogy, and physical properties* (pp. 599–623). Cambridge University Press.

Devitt, M. 2010. Species have (partly) intrinsic essences. *Philosophy of Science* 77: 648–661.

Devolder, K. 2019. Embryo research. In D. Edmonds (Ed.). *Ethics and the contemporary world* (pp. 249–264). Routledge, Taylor & Francis Group.

de Lazari-Radek, K., and P. Singer. 2017. *Utilitarianism: a very short introduction*. Oxford University Press.

De Wert, G., Heindryckx B, Pennings G, et al., the European Society of Human Genetics and the European Society of Human Reproduction and Embryology. 2018a. Responsible innovation in human germline gene editing: Background document to the recommendations of ESHG and ESHRE. *European Journal of Human Genetics* 26 (4): 450–470.

De Wert, G., G. Pennings, A. Clarke, U. et al., the European Society of Human Genetics and the European Society of Human Reproduction and Embryology. 2018b. Human germline gene editing: Recommendations of ESHG and ESHRE. *Human Reproduction Open* 1–5. https://doi.org/10.1093/hropen/hox025.

Dick, S. J. 1998. *Life on other worlds: The 20th-century extraterrestrial life debate*. Cambridge University Press.

Douglas, T. 2019. Genetic selection. In D. Edmonds (Ed.)., *Ethics and the contemporary world* (pp. 303–317). Routledge, Taylor & Francis Group.

Edwards, M. R. 2021. Android Noahs and embryo arks: Ectogenesis in global catastrophe survival and space colonization. *International Journal of Astrobiology* 20: 150–158.

Elliott, C. 2004. Pharma's gain may be our loss. *PLOS Medicine* 1 (3): 52–53.

Elvis, M. 2021. *Asteroids: How love, fear, and greed will determine our future in space.* Yale University Press.

Elvis, M., and T. Milligan. 2019. How much of the Solar System should we leave as wilderness? *Acta Astronautica* 162: 574–580.

Erler, A. 2020. Enriching, rather than revising: The conceptual toolbox on germline interventions. *American Journal of Bioethics* 20 (8): 25–27.

Erler, A., and V. C. Müller. 2021. The ethics of biomedical military research: Therapy, prevention, enhancement, and risk. In D. Messelken and D. Winkler (Eds.), *Health care in contexts of risk, uncertainty, and hybridity* (pp. 235–252). Springer.

Evans, J. H. 2020. *The human gene editing debate.* Oxford University Press.

Fanciullo, J. 2020. Human enhancement and the proper response to climate change. *Ethics, Policy and Environment* 23 (1): 85–96.

Feeney, O., and V. Rakić. 2021. Genome editing and "disenhancement": Considerations on issues of non-identity and genetic pluralism. *Humanities and Social Sciences Communications* 8: 116. https://doi.org/10.1057/s41599-021-00795-w.

Flexner, A. 2017. *The usefulness of useless knowledge.* Princeton University Press.

Forget, F., F. Costard, and P. Lognonné. 2006. *Planet Mars: Story of another world.* Springer, Praxis Publishing.

Futures. 2019. K. C. Smith, and K. Abney (Eds.). Human colonization of other worlds. *Futures* 110: 1–66.

Futuyma, D. J. 2006. *Evolution.* Sinauer Associates.

Garrett-Bakelman, F. E., M. Darshi, S. J. Green, et al. 2019. The NASA twins study: A multidimensional analysis of a year-long human spaceflight. *Science* 364, eaau8650.

Gaskell, G., I. Bard, A. Allansdottir, et al. 2017. Public views on gene editing and its uses. *Nature Biotechnology* 35 (11): 1021–1023.

Gheaus, A. 2016. The right to parent and duties concerning future generations. *Journal of Political Philosophy* 24: 487–508.

Gibson, D. C. 2012. *Commercial space tourism: Impediments to industrial development and strategic communication solutions.* Bentham eBooks.

Gibson, R. B. 2021. The epidemiology of moral bioenhancement. *Medicine, Health Care, and Philosophy* 24 (1): 45–54.

Gidley, J. M. 2017. *The future: A very short introduction.* Oxford University Press.

Gilligan, C. 1982. *In a different voice.* Harvard University Press.

Giubilini, A., and S. Sanyal. 2015. The ethics of human enhancement. *Philosophy Compass* 10 (4): 233–243.

Giubilini, A., and S. Sanyal. 2016. Challenging human enhancement. In S. Clarke, J. Savulescu, C. A. J. Coady, A. Giubilini, and S. Sanyal. 2016. *The ethics of human enhancement: Understanding the debate* (pp. 1–24). Oxford University Press.

Glannon, W. 2001. *Genes and future people: Philosophical issues in human genetics.* Westview Press.

Goldsmith, D., and M. Rees. 2022. *The end of astronauts: Why robots are the future of exploration.* Belknap Press of Harvard University Press.

Gould, D. C. 2021. Future minds and a new challenge to anti-natalism. *Bioethics* 35: 793–800, https://doi.org/10.1111/bioe.12873.

Gowdy, J. 2020. Our hunter-gatherer future: Climate change, agriculture and uncivilization. *Futures* 115: 102488.

Grant, D. 2021. *Privacy in the age of neuroscience: Reimagining law, state and market.* Cambridge University Press.

Gray, P. B., and J. R. Garcia. 2013. *Evolution and human sexual behavior.* Harvard University Press.

Greely, H. T. 2021. *CRISPR people: The science and ethics of editing humans.* MIT Press.

Green, B. P. 2020. Convergences in the ethics of space exploration. In K. C. Smith and C. Mariscal (Eds.), *Social and conceptual issues in astrobiology* (pp. 179–196). Oxford University Press.

Green, B. P. 2021. *Space ethics.* Rowman & Littlefield.

Greenbaum, D., and L.Y. Cabrera. 2020. Editorial: ELSI in human enhancement: What distinguishes it from therapy? *Frontiers in Genetics* 11: 618: 1–3, https://doi.org/10.3389/fgene.2020.00618.

Greene, M., and Z. Master. 2018. Ethical issues of using CRISPR technologies for research on military enhancement. *Journal of Bioethical Inquiry* 15: 327–335.

Grego, L. 2021. Outer space and crisis risk. In C. Steer and M. Hersch (Eds.), *War and peace in outer space* (pp. 265–285). Oxford University Press.

Gross, M. L. 2006. *Bioethics and armed conflict: Moral dilemmas of medicine and war.* MIT Press.

Gunderson, R., D. Stuart, and B. Petersen. 2021. In search of plan(et) B: Irrational rationality, capitalist realism, and space colonization. *Futures* 134: 102857.

Gyngell, C. 2017. Gene editing and the health of future generations. *Journal of the Royal Society of Medicine* 110 (7): 276–279.

Gyngell, C., H. Bowman-Smart, and J. Savulescu. 2019. Moral reasons to edit the human genome: Picking up from the Nuffield report. *Journal of Medical Ethics* 45: 514–523.

Gyngell, C., T. Douglas, and J. Savulescu. 2017. The ethics of germline gene editing. *Journal of Applied Philosophy* 34 (4): 498–513.

Gyngell, C., and M. J. Selgelid. 2016. Human enhancement: Conceptual clarity and moral significance. In S. Clarke, J. Savulescu, C. A. J. Coady, A. Giubilini, and S. Sanyal. 2016. *The ethics of human enhancement: Understanding the debate* (pp. 111–126). Oxford University Press.

Haqq-Misra, J. 2019. Can deep altruism sustain space settlement? In K. Szocik (ed.), *The human factor in a mission to Mars: An interdisciplinary approach* (pp. 145–155). Springer.

Harris, J. 2007. *Enhancing evolution: The ethical case for making better people.* Princeton University Press.

Harris, J. 2011. Enhancements are a moral obligation. In J. Savulescu and N. Bostrom (Eds.), *Human enhancement* (pp. 131–154). Oxford University Press.

Harris, J. 2011. Moral enhancement and freedom. *Bioethics* 25 (2): 102–111.

Hauskeller, M. 2013. *Better humans? Understanding the enhancement project.* Routledge.

Hauskeller, M. 2016. Levelling the playing field on the alleged unfairness of the genetic lottery. In S. Clarke, J. Savulescu, C. A. J. Coady, A. Giubilini, and S. Sanyal (Eds.), *The ethics of human enhancement: Understanding the debate* (pp. 199–210). Oxford University Press.

Hauskeller, M. 2019. Editing the best of all possible worlds. In E. Parens and J. Johnston (Eds.), *Human flourishing in an age of gene editing* (pp. 61–71). Oxford University Press.

Hertzfeld, H. 2009. Current and future issues in international space law. *ILSA Journal of International & Comparative Law* 15 (2): 325–336.

Hertzfeld, H. R. 2021. Unsolved issues of compliance with the registration convention. *Journal of Space Safety Engineering* 8 (3): 238–244.

Hertzfeld, H. R., and S. N. Pace. 2013. International cooperation on human lunar heritage. *Science* 342 (6162): 1049–1050.

Hertzfeld, H. R., B. Weeden, and C. D. Johnson. 2016. Outer space: Ungoverned or lacking effective governance? New approaches to managing human activities in space. *SAIS Review of International Affairs* 36 (2): 15–28.

Huang, K., J. D. Greene, and M. Bazerman. 2019. Veil-of-ignorance reasoning favors the greater good. *Proceedings of the National Academy of Sciences* 116 (48): 23989–23995.

Hull, D. L. 1986. On human nature. *PSA: Proceedings of the Biennial Meeting of the Philosophy of Science Association. Volume Two: Symposia and Invited Papers*, A. Woody (ed.) (pp. 3–13). The University of Chicago Press.

Impey, C. 2015. *Beyond: Our future in space*. Norton.

Impey, C. 2021. Science and faith off-Earth. In M. Boone Rappaport and K. Szocik (Eds.), *The human factor in the settlement of the Moon: An interdisciplinary approach* (pp. 245–255). Springer.

Johns Hopkins University. 2021. COVID-19 dashboard by the Center for Systems Science and Engineering (CSSE) at Johns Hopkins University (JHU). Coronavirus Resource Center. https://coronavirus.jhu.edu/map.html.

Kamm, F. 2009. What is and is not wrong with enhancement? In J. Savulescu and N. Bostrom (Eds.), *Human enhancement* (pp. 91–130). Oxford University Press.

Kargel, J. S. 2004. *Mars—A warmer, wetter planet*. Springer Verlag.

Keown, J. 1997. *Euthanasia examined: Ethical, clinical and legal perspectives*. Cambridge University Press.

Kessler, D. J., and B. G. Cour-Palais. 1978. Collision frequency of artificial satellites: The creation of a debris belt. *Journal of Geophysical Research* 83 (A6): 2637–2646.

Keszthelyi, L., et al. 2017. *Feasibility study for the quantitative assessment of mineral resources in asteroids*. U.S. Department of the Interior, U.S. Geological Survey.

Kieffer, H. H., B. M. Jakosky, C. W. Snyder, and M. S. Matthews (Eds.). 1992. *Mars*. University of Arizona Press.

Kirksey, E. 2020. *The mutant project: Inside the global race to genetically modify humans*. La Trobe University Press, in conjunction with Black Inc.

Kleiderman, E., and U. Ogbogu. 2019. Realigning gene editing with clinical research ethics: What the "CRISPR twins" debacle means for Chinese and international research ethics governance. *Accountability in Research* 26 (4): 257–264.

Klitzman, R. L. 2020. *Designing babies: How technology is changing the ways we create children*. Oxford University Press.

Klug, W. S., M. R. Cummings, C. A. Spencer, M. A. Palladino, and D. J. Killian. 2020. *Essentials of genetics* (10th ed.). Pearson.

Knowledgemotion. 2018. *How asteroid mining will save Earth*. Knowledgemotion. Distributed by Infobase, 2020.

Koepsell, D. 2017. Mars One: Human subject concerns? *Astropolitics* 15 (1): 97–111.

Kolk, M. 2022. Demographic theory and population ethics. In G. Arrhenius, K. Bykvist, T. Campbell, and E. Finneron-Burns (Eds.), *The Oxford handbook of population ethics* (pp. 468–490). Oxford University Press.

Kovic, M. 2021. Risks of space colonization, *Futures* 126: 102638.

Krag, H. 2021. A sustainable use of space. *Science* 373 (6552): 259.

Kramer, W. R. 2020. A framework for extraterrestrial environmental assessment. *Space Policy* 53: 101385.

Krimsky, S. 2019. Ten ways in which He Jiankui violated ethics. *Nature Biotechnology* 37 (1): 19–20.

Kudlek, K. 2021. Is human enhancement intrinsically bad? *Medicine, Health Care and Philosophy* 24: 269–279.

Kudlek, K. 2022. On the uneasy alliance between moral bioenhancement and utilitarianism. *Bioethics* 36: 210–217, https://doi.org/10.1111/bioe.12974.

Kuhse, H., and P. Singer. 2009. What is bioethics? A historical introduction. In H. Kuhse and P. Singer (Eds.), *A companion to bioethics* (2nd ed., pp. 3–11). Wiley-Blackwell.

Kurtz, S. A. 2021. Beyond the lifetime of organizations: A framework for multi-generational goal survival in the ecology of goals. *Futures* 127: 102699.

Latheef, S., and A. Henschke. 2020. Can a soldier say no to an enhancing intervention? *Philosophies* 5 (3): 13: 1–17, https://doi.org/10.3390/philosophies5030013.

Le Dévédec's, N. 2020. The biopolitical embodiment of work in the era of human enhancement. *Body and Society* 26 (1): 55–81.

Lee, C. L., K. Castle, E. Moding, et al. 2015. Acute DNA damage activates the tumour suppressor p53 to promote radiation-induced lymphoma. *Nature Communications* 6: 8477: 1–12.

Leslie, J. 1996. *The end of the world: The science and ethics of human extinction.* Routledge.

Lewens, T. 2015. *The biological foundations of bioethics.* Oxford University Press.

Limaye, S. S., R. Mogul, K. H. Baines, et al. 2021. Venus, an astrobiology target. *Astrobiology* 21 (8): 1163–1185.

Łukasiewicz, J. 1961. Uwagi filozoficzne o wielowartościowych systemach rachunku zdań. In J. Łukasiewicz, *Z zagadnień logiki i filozofii. Pisma wybrane* (pp. 144–163). Państwowe Wydawnictwo Naukowe.

Luo, Y (Ed.). 2019. *CRISPR gene editing: Methods and protocols.* Humana Press.

Ma, H., et al. 2017. Correction of a pathogenic gene mutation in human embryos. *Nature* 548: 413–419.

Machery, E. 2008. A plea for human nature. *Philosophical Psychology* 21: 321–329.

Magni, S. F. 2021. In defence of person-affecting procreative beneficence. *Bioethics* 35: 473–479, https://doi.org/10.1111/bioe.12872.

Manning, R. C. 2009. A care approach. In H. Kuhse and P. Singer (Eds.), *A companion to bioethics* (2nd ed., pp. 105–116). Wiley-Blackwell.

Marshall, W. 2001. Limited by cost: The case against humans in the scientific exploration of space—Discussion. *Earth, Moon, and Planets* 87 (3): 218–219.

Mason, C. E. 2021. *The next 500 years: Engineering life to reach new worlds.* MIT Press.

Mazur, S. K., D. Minich, and K. Szocik. 2020. Legal, political, and ethical challenges of Mars settlement. *Astropolitics* 18 (3): 223–237.

McCormick, H. 2009. Intergenerational justice and the nonreciprocity problem. *Political Studies* 57 (2): 451–58.

McKay, C. P. 1990. Does Mars have rights? An approach to the environmental ethics of planetary engineering, In D. MacNiven (Ed.), *Moral expertise: Studies in practical and professional ethics* (pp. 184–197). Routledge.

McMillan, J. 2018. *The methods of bioethics: An essay in meta-bioethics.* Oxford University Press.

Memi, F., A. Ntokou, and I. Papangeli. 2018. CRISPR/Cas9 gene-editing: Research technologies, clinical applications and ethical considerations. *Seminars in Perinatology* 42: 487–500.

Meneely, P. 2020. *Genetic analysis: Genes, genomes, and networks in eukaryotes* (3rd ed.). Oxford University Press.

Mepham, T. B. 2008. *Bioethics: An introduction for the biosciences.* Oxford University Press.

Messelken, D., and D. Winkler (Eds.). 2020. *Ethics of medical innovation, experimentation, and enhancement in military and humanitarian contexts.* Springer.

Metzger, P. T. 2016. Space development and space science together, an historic opportunity. *Space Policy* 37 (Part 2): 77–91.

Miller, G. 2000. *The mating mind: How sexual choice shaped the evolution of human nature.* Anchor Books.

Milligan, T. 2011. Property rights and the duty to extend human life. *Space Policy* 27: 190–193.

Milligan, T. 2015a. Asteroid mining, integrity and containment. In J. Galliott (Ed.), *Commercial space exploration: Ethics, policy and governance* (pp. 123–134). Ashgate.

Milligan, T. 2015b. *Nobody owns the Moon: The ethics of space exploitation.* McFarland.

Milligan, T. 2016. Common origins and the ethics of planetary seeding. *International Journal of Astrobiology* 15 (4): 301–306.

Morrison, D., C. R. Chapman, D. Steel, and R. P. Binzel. 2004. Impacts and the public: Communicating the nature of the impact hazard. In M. J. S. Belton, T. H. Morgan,

N. H. Samarasinha, and D, K. Yeomans (Eds.), *Mitigation of hazardous comets and asteroids* (pp. 353–390). Cambridge University Press.

Mulgan, T. 2006. *Future people: A modern consequentialist account of our obligations to future generations*. Clarendon Press.

Munévar, G. 2023. *The dimming of starlight: The philosophy of space exploration*. Oxford University Press.

NASA. 2021. Artificial gravity. National Aeronautics and Space Administration, March 26, 2021. https://www.nasa.gov/johnson/HWHAP/artificial-gravity.

National Council on Radiation Protection and Measurements. 2011. *Potential impact of individual genetic susceptibility and previous radiation exposure on radiation risk for astronauts: Recommendations of the National Council on Radiation Protection and Measurements, April 28, 2010*. National Council on Radiation Protection and Measurements, Bethesda, Maryland.

National Research Council (U.S.). 2005. *Science in NASA's vision for space exploration*. National Research Council (U.S.). Committee on the Scientific Context for Space Exploration. National Academies Press.

Navarro, V. (Ed.). 1992. *Why the United States does not have a national health program*. Baywood Publishing.

Nettleton, S. 1995. *The sociology of health and illness*. Polity Press.

Newhouse, J. P., and The Insurance Experiment Group. 1993. *Free for all? Lessons from the RAND health insurance experiment; A RAND study*. Harvard University Press.

Newman, S. 2000. Epigenetic vs. genetic determinism,. In C. Walker (Ed.), *Made not born: The troubling world of biotechnology* (pp. 27–45). Sierra Club Books.

NHGRI. n.d. *Germ line*. National Human Genome Research Institute (NHGRI), https://www.genome.gov/glossary/index.cfm?id=94.

Norenzayan, A. 2013. *Big gods: How religion transformed cooperation and conflict*. Princeton University Press.

Norman, Z., and M. J. Reiss. 2021. The emergence of an environmental ethos on Luna. In M. Boone Rappaport and K. Szocik (Eds.), *The human factor in the settlement of the Moon: An interdisciplinary approach* (pp. 221–232). Springer.

Norris, P., and R. Inglehart. 2004. *Sacred and secular: Religion and politics worldwide*. Cambridge University Press.

Ord, T. 2020. *The precipice: Existential risk and the future of humanity*. Bloomsbury Publishing.

Osinski, G. R., C. S. Cockell, A. Pontefract, and H. M. Sapers. 2020. The role of meteorite impacts in the origin of life. *Astrobiology* 20 (9): 1121–1149.

Overall, C. 2009. Life enhancement technologies: The significance of social category membership. In J. Savulescu and N. Bostrom (Eds.), *Human enhancement* (pp. 327–340). Oxford University Press.

Palacios-González, C. 2021. Reproductive genome editing interventions are therapeutic, sometimes. *Bioethics* 35: 557–562. https://doi.org/10.1111/bioe.12846.

Paladini, S. 2019. *The new frontiers of space: Economic implications, security issues and evolving scenarios*. Springer International Publishing.

Palmroth, M., J. Tapio, A. Soucek, et al. 2021. Toward sustainable use of space: Economic, technological, and legal perspectives. *Space Policy* 57: 101428.

Parfit, D. 1984. *Reasons and persons*. Clarendon Press.

Parfit, D. 2017. Future people, the non-identity problem, and person-affecting principles. *Philosophy and Public Affairs* 45 (2): 118–157.

Pavarini, G., R. McMillan, A. Robinson, and I. Singh. 2021. Design bioethics: A theoretical framework and argument for innovation in bioethics research. *American Journal of Bioethics* 6: 37–50.

Pelton, J. N. 2017. *The new gold rush: The riches of space beckon!* Springer International Publishing.

Peters, T. 2019a. Are we closer to free market eugenics? The CRISPR controversy. *Zygon* 54 (1): 7–13.

Peters, T. 2019b. Does extraterrestrial life have intrinsic value? An exploration in responsibility ethics. *International Journal of Astrobiology* 18 (4): 304–310.

Peters, T. 2021. Astroethics for earthlings: Our responsibility to the galactic commons. In O. A. Chon-Torres, T. Peters, J. Seckbach, and R. Gordon (Eds.), *Astrobiology: Science, ethics, and public policy* (pp. 17–56). Wiley-Scrivener Publishing.

Peterson, M. 2017. *The ethics of technology: A geometric analysis of five moral principles.* Oxford University Press.

Pilchman, D. 2015. Three ethical perspectives on asteroid mining. In J. Galliott (Ed.), *Commercial space exploration: Ethics, policy and governance* (pp. 123–134). Ashgate.

Plomin, R. 2018. *Blueprint: How DNA makes us who we are.* Allen Lane, an imprint of Penguin Books.

Plomin, R., and S. von Stumm. 2018. The new genetics of intelligence. *Nature Reviews Genetics* 19 (3): 148–159.

Powell, R. 2015. In genes we trust: Germline engineering, eugenics, and the future of the human genome. *Journal of Medicine and Philosophy* 40: 669–695.

Powell, R., and A. Buchanan. 2016. The evolution of moral enhancement. In S. Clarke, J. Savulescu, C. A. J. Coady, A. Giubilini, and S. Sanyal (Eds.), *The ethics of human enhancement: Understanding the debate* (pp. 239–260). Oxford University Press.

President's Council on Bioethics. 2002. *Staff background paper: Human genetic enhancement.*

President's Council on Bioethics. 2003. *Beyond therapy: Biotechnology and the pursuit of happiness; A report of the President's Council on Bioethics.* October 2003. https://biotech.law.lsu.edu/research/pbc/reports/beyondtherapy/.

Pugh, J., G. Kahane, J. Savulescu. 2013. Cohen's conservatism and human enhancement. *Ethics* 17: 331–354.

Pullman, D. 1999. The ethics of autonomy and dignity in long-term care. *Canadian Journal on Aging* 18 (1): 26–46.

Rakić, V. 2021a. *How to enhance morality.* Springer International Publishing.

Rakić, V. 2021b. *The ultimate enhancement of morality.* Springer International Publishing.

Randall, T. 2019. Care ethics and obligations to future generations. *Hypatia* 34 (3): 527–545.

Rappaport, M. B., C. Corbally, and K. Szocik. 2021. Interstellar ethics and the Goldilocks evolutionary sequence: Can we expect ETI to be moral? In O. A. Chon-Torres, T. Peters, J. Seckbach, and R. Gordon (Eds.). *Astrobiology: Science, ethics, and public policy* (pp. 339–359). Wiley-Scrivener Publishing.

Rappaport, M. B., and K. Szocik (Eds.). 2021a. *The human factor in the settlement of the Moon: An interdisciplinary approach.* Springer.

Rappaport, M. B., and K. Szocik. 2021b. Practical planning commences: Next Steps in the settlement of Earth's Moon. In M. B. Rappaport and K. Szocik (Eds.), *The human factor in the settlement of the Moon. An interdisciplinary approach* (pp. 1–16). Springer.

Ravitsky, V. 2021. *Ethical and social aspects of germline gene editing.* Presentation given at an online meeting of the Yale Technology and Ethics group, October 14, 2021.

Rawls, J. 1999. *A theory of justice.* Belknap Press.

Rees, M. J. 2003. *Our final hour: A scientist's warning; How terror, error, and environmental disaster threaten humankind's future in this century—on Earth and beyond.* Basic Books.

Richerson, P. J., S. Gavrilets, and F. B. M. de Waal. 2021. Modern theories of human evolution foreshadowed by Darwin's *Descent of Man. Science* 372 (6544), eaba3776. https://doi.org/10.1126/science.aba3776.

Roache, R., and J. Savulescu. 2016. Enhancing conservatism. In S. Clarke, J. Savulescu, C. A. J. Coady, A. Giubilini, and S. Sanyal (Eds.), *The ethics of human enhancement: Understanding the debate* (pp. 145–159). Oxford University Press.

Robinson, W. M. 2004. Ethics for astronauts. *Medical Ethics* 11 (3): 1–2.

Roduit, J. A. R., J.-C. Heilinger, and H. Baumann. 2015. Ideas of perfection and the ethics of human enhancement. *Bioethics* 29 (9): 622–630.

Ross, C. T., and P. J. Richerson. 2014. New frontiers in the study of human cultural and genetic evolution. *Current Opinion in Genetics and Development* 29: 103–109.

Rothman, S. 2015. *The paradox of evolution: The strange relationship between natural selection and reproduction.* Prometheus Books.

Rubin, C. T. 2014. *Eclipse of man: Human extinction and the meaning of progress.* New Atlantis Books.

Rueda, J. 2021. Climate change, moral bioenhancement and the ultimate mostropic. *Ramon Llull Journal of Applied Ethics* 11 (11): 277–303.

Rueda, J., P. García-Barranquero, and F. Lara. 2021. Doctor, please make me freer: Capabilities enhancement as a goal of medicine. *Medicine, Health Care and Philosophy* 24: 409–419.

Rulli, T. 2019. Reproductive CRISPR does not cure disease. *Bioethics* 33 (9): 1072–1082. https:// doi.org/10.1111/bioe.12663.

Sadeh, E. (Ed.) 2013. *Space strategy in the 21st century: Theory and policy.* Routledge.

Sandel, M. 2004. *The case against perfection.* The Atlantic. https://www.theatlantic.com/magaz ine/archive/2004/04/the-case-against-perfection/302927/.

Savulescu, J. 2001a. In defense of selection for non-disease genes. *American Journal of Bioethics* 1 (1): 16–19.

Savulescu, J. 2001b. Procreative beneficence: Why we should select the best children. *Bioethics* 15 (5–6): 413–426.

Savulescu, J. 2019. Human enhancement. In D. Edmonds (Ed.), *Ethics and the contemporary world* (pp. 319–334). Routledge, Taylor & Francis Group.

Savulescu, J, C. Gyngell, and G. Kahane. 2021. Collective reflective equilibrium in practice (CREP) and controversial novel technologies. *Bioethics* 35: 652–663. https://doi.org/ 10.1111/bioe.12869.

Savulescu, J., and G. Kahane. 2009. The moral obligation to create children with the best chance of the best life. *Bioethics* 23 (5): 274–290.

Savulescu, J., B. Foddy, and M. Clayton. 2004. Why we should allow performance enhancing drugs in sport. *British Journal of Sports Medicine* 38: 666–670.

Savulescu, J., Sandberg, A., and Kahane, G. 2011. Well-being and enhancement. In J. Savulescu, R. ter Meulen, and G. Kahane (Eds.), *Enhancing human capacities* (pp. 3–18). Blackwell.

Sawin, C. 2017. Bioethics in space exploration. In L. Young and J. Young (Eds.), *Encyclopedia of bioastronautics* (pp. 1–6). Springer. https://doi.org/10.1007/978-3-319-10152-1_136-2.

Sawin, C. 2021. Bioethics in space exploration. In L. R. Young and J. P. Sutton (Eds.), *Handbook of bioastronautics* (pp. 565–572). Springer.

Schleim, S., and B. B. Quednow. 2018. How realistic are the scientific assumptions of the neuroenhancement debate? Assessing the pharmacological optimism and neuroenhancement prevalence hypotheses. *Frontiers in Pharmacology* 9: 3.

Schmidt, N., and P. Bohacek. 2021. First space colony: What political system could we expect? *Space Policy* 56: 101426.

Schneider, S. 2019. *Artificial you: AI and the future of your mind.* Princeton University Press.

Schulze-Makuch, D., R. Heller, and E. Guinan. 2020. In search for a planet better than Earth: Top contenders for a superhabitable world. *Astrobiology* 20 (12): 1394–1404.

Schwartz, J. S. J. 2018. Worldship ethics: Obligations to the crew. *Journal of the British Interplanetary Society* 71: 53–64.

Schwartz, J. S. J. 2020. The accessible universe: On the choice to require bodily modification for space exploration. In K. Szocik (Ed.), *Human enhancements for space missions: Lunar, martian, and future missions to the outer planets* (pp. 201–215). Space and Society. Springer.

Schwartz, J. S. J. 2020. *The value of science in space exploration.* Oxford University Press.

Schwartz, J. S. J. 2021. A Right to Return to Earth? Emigration Policy for the Lunar State. In M. Boone Rappaport and K. Szocik (Eds.), *The human factor in the settlement of the Moon: An interdisciplinary approach* (pp. 193–205). Springer.

Schwartz, J. S. J., and T. Milligan (Eds.). 2016. *The ethics of space exploration.* Springer.

Schwartz, J. S. J., and T. Milligan. 2016. Introduction: The scope and content of space ethics. In J. S. J. Schwartz and T. Milligan (Eds.), *The ethics of space exploration* (pp. 1–11). Springer.

Schwartz, J. S. J., S. Wells-Jensen, J. W. Traphagan, D. Weibel, and K. Smith. 2021. What do we need to ask before settling space? *Journal of the British Interplanetary Society* 74: 2–9.

Segovia-Cuéllar, A., and L. Del Savio. 2021. On the use of evolutionary mismatch theories in debating human prosociality. *Medicine, Health Care, and Philosophy* 24 (3): 305–314.

Shelhamer, M. 2017. Why send humans into space? Science and non-science motivations for human space flight. *Space Policy* 42: 37–40.

Shook, J. R., and J. J. Giordano. 2017. Moral bioenhancement for social welfare: Are civic institutions ready? *Frontiers in Sociology* 2: 21.

Simkulet, W. 2021. Genetic parenthood and hard cases. *Bioethics* 35: 680–687.

Singer, P. 2009. Parental choice and human improvement. In J. Savulescu and N. Bostrom (Eds.), *Human enhancement* (pp. 277–289). Oxford University Press.

Sivolella, D. 2019. *Space mining and manufacturing: Off-world resources and revolutionary engineering techniques.* Springer International Publishing.

Slaba, T. C., et al. 2017. Optimal shielding thickness for galactic cosmic ray environments. *Life Sciences in Space Research* 12: 1–15.

Smith, K. C. 2016. The curious case of the martian microbes: Mariomania, intrinsic value and the prime directive. In J. S. J. Schwartz and T. Milligan (Eds.), *The ethics of space exploration* (pp. 195–208). Springer.

Smith, K. C., and C. Mariscal. 2020. To the humanities and beyond: Exploring the broader questions in astrobiology. In K. C. Smith and C. Mariscal (Eds.), *Social and conceptual issues in astrobiology* (pp. 3–6). Oxford University Press.

Smith, K. C., and C. Mariscal (Eds.). 2020. *Social and conceptual issues in astrobiology.* Oxford University Press.

Smith, K. C., and J. W. Traphagan. 2020. First, do nothing: A passive protocol for first contact. *Space Policy* 54: 101389.

Sniekers, S., S. Stringer, K. Watanabe, et al. 2017. Genome-wide association meta-analysis of 78,308 individuals identifies new loci and genes influencing human intelligence. *Nature Genetics* 49: 1107–1112.

Snyder-Beattie, A. E., A. Sandberg, K. Eric Drexler, and M. B. Bonsall. 2021. The timing of evolutionary transitions suggests intelligent life is rare. *Astrobiology* 21 (3): 265–278.

Sober, E. 1993. *Philosophy of biology.* Westview.

Sparrow, R. 2011. A Not-So-New Eugenics. *Hastings Center Report* 41: 32–42.

Sparrow, R. 2014. (Im)moral technology? Thought experiments and the future of "mind control." In A. Akabayashi (Ed.), *The future of bioethics: International dialogues* (pp. 113–119). Oxford University Press.

Sparrow, R. 2016. Human enhancement for whom? In S. Clarke, J. Savulescu, C. A. J. Coady, A. Giubilini, and S. Sanyal (Eds.), *The ethics of human enhancement: Understanding the debate* (pp. 127–142). Oxford University Press.

Sparrow, R. 2019. Yesterday's child: How gene editing for enhancement will produce obsolescence—and why it matters. *American Journal of Bioethics* 19 (7): 6–15.

Sparrow, R. 2021. Human germline genome editing: On the nature of our reasons to genome edit. *American Journal of Bioethics* 22 (9): 4–15. https://doi.org/10.1080/15265161.2021.1907480.

Spector, S. 2020. Delineating acceptable risk in the space tourism industry. *Tourism Recreation Research* 45 (4): 500–510.

Steer, C., and M. Hersch. 2021. Conclusion. In C. Steer and M. Hersch (Eds.), *War and peace in outer space* (pp. 301–308). Oxford University Press.

Sterelny, K., and P. Griffiths. 1999. *Sex and death*. University of Chicago Press.

Stock, G. 2002. *Redesigning humans: Our inevitable genetic future*. Houghton Mifflin Company.

Szocik, K. 2015. Mars, human nature and the evolution of the psyche. *Journal of the British Interplanetary Society* 68 (12): 403–405.

Szocik, K. 2019a. Biomedical moral enhancement for human space missions. *Studia Humana* 8 (4): 1-9.

Szocik, K. (Ed.). 2019b. *The human factor in a mission to Mars: An interdisciplinary approach*. Springer.

Szocik, K. 2019c. Human place in the outer space: Skeptical remarks. In K. Szocik (Ed.), *The human factor in a mission to Mars: An interdisciplinary approach* (pp. 233–252). Springer.

Szocik, K. 2019d. Should and could humans go to Mars? Yes, but not now and not in the near future. *Futures* 105: 54–66..

Szocik, K. 2019e. What is right and what is wrong in the Darwinian approach to the study of religion. *Social Evolution and History* 18 (2): 210–228.

Szocik, K. 2020a. Human enhancement and Mars settlement—Biological necessity or science-fiction? The special case of biomedical moral enhancement for future space missions. In K. Szocik (Ed.), *Human enhancements for space missions: Lunar, martian, and future missions to the outer planets* (pp. 253–264). Springer.

Szocik, K. (Ed.). 2020b. *Human enhancements for space missions: Lunar, martian, and future missions to the outer planets*. Springer.

Szocik, K. 2020c. Human future in space and gene editing: Waiting for feminist space ethics and feminist space philosophy. *Theology and Science* 18 (1): 7–10.

Szocik, K. 2020d. Is human enhancement in space a moral duty? Missions to Mars, advanced AI and genome editing in space. *Cambridge Quarterly of Healthcare Ethics* 29 (1): 122–130.

Szocik, K. 2021a. The biologically optimized spacefarer. *Science* 372 (6541): 469.

Szocik, K. 2021b. Ethical, political and legal challenges relating to colonizing and terraforming Mars. In M. Beech, J. Seckbach, and R. Gordon. *Terraforming Mars* (pp. 123–134). Astrobiology Perspectives on Life of the Universe. Scrivener Publishing—Wiley.

Szocik, K. 2021c. Humanity should colonize space in order to survive but not with embryo space colonization. *International Journal of Astrobiology* 20 (4): 319–322.

Szocik, K. 2021d. *List of publications*, Wyższa Szkoła Informatyki i Zarządzania z siedzibą w Rzeszowie. https://wsiz.edu.pl/kadra-akademicka/kszocik/.

Szocik, K. 2021e. Space bioethics: Why we need it and why it should be a feminist space bioethics. *Bioethics* 35 (2): 187–191.

Szocik, K. 2021f. Lunar settlement, space refuge, and quality of life: A prevention policy for the future of humans on Luna. In M. Boone Rappaport and K. Szocik (Eds.), *The human factor in the settlement of the Moon. An interdisciplinary approach* (pp. 209–220). Springer.

Szocik, K. 2023. Ethical issues in CRISPR for human space missions. In T. Peters and A. Gouw (Eds.), *The CRISPR Revolution in Science, Religion, and Ethics*, 1 volume. Praeger.

Szocik, K. 2022. Evolutionary biology as a source of reliable knowledge about extraterrestrial intelligence (ETI): Why we should reject militarism in our thinking about ETI. In J. Andresen and O. A. Chon Torres (Eds.), *Extraterrestrial intelligence: Academic and societal implications* (pp. 173–186). Cambridge Scholars Publishing.

Szocik, K., Abood S., Impey C., et al. 2020. Visions of a martian future. *Futures* 117, 102514.

Szocik, K., S. Abood, and M. Shelhamer. 2018. Psychological and biological challenges of the Mars mission viewed through the construct of the evolution of fundamental human needs. *Acta Astronautica* 152: 793–799.

Szocik, K., and R. Abylkasymova. 2022a. Covid-19 pandemic and future global catastrophic risks as a challenge for health-care ethics. *International Journal of Human Rights in Healthcare* 15 (4): 340–350. https://doi.org/10.1108/IJHRH-12-2020-0107.

Szocik, K., and R. Abylkasymova. 2022b. Ethical issues in police robots: The case of crowd control robots in a pandemic. *Journal of Applied Security Research* 17 (4): 530–545. https://doi.org/10.1080/19361610.2021.1923365.

Szocik, K., and M. Braddock. 2019. Why human enhancement is necessary for successful human deep-space missions. *New Bioethics* 25 (4): 295–317.

Szocik, K., R. Campa, M. B. Rappaport, and C. Corbally. 2019. Changing the paradigm on human enhancements: The special case of modifications to counter bone loss for manned Mars missions. *Space Policy* 48: 68–75.

Szocik, K., R. Elias Marques, S. Abood, K. Lysenko-Ryba, D. Minich, and A. Kędzior. 2018. Biological and social challenges of human reproduction in a long-term Mars base. *Futures* 100: 56–62.

Szocik, K., and A. M. Gouw. 2023. Moral bioenhancement for space: Should we enhance morally future deep-space astronauts and space settlers? *Theology and Science*.

Szocik, K., K. Lysenko-Ryba, S. Banaś, and S. Mazur. 2016. Political and legal challenges in a Mars colony. *Space Policy* 38: 27–29.

Szocik, K., Z. Norman, and M. J. Reiss. 2020. Ethical challenges in human space missions: A space refuge, scientific value, and human gene editing for space. *Science and Engineering Ethics* 26: 1209–1227.

Szocik, K., M. Shelhamer, M. Braddock, et al. 2021. Future space missions and human enhancement: Medical and ethical challenges. *Futures* 133: 102819.

Szocik, K., and K. Tachibana. 2019. Human enhancement and artificial intelligence for space missions. *Astropolitics* 17 (3): 208–219.

Szocik, K., and B. Tkacz. 2018. Multi-level challenges in a long-term human space program: The case of manned mission to Mars. *Studia Humana* 7 (2): 24–30.

Szocik, K., T. Wójtowicz, and L. Baran. 2017. War or peace? The possible scenarios of colonising Mars. *Space Policy* 42: 31–36.

Szocik, K., T. Wójtowicz, M. B. Rappaport, and C. Corbally. 2020. Ethical issues of human enhancements for space missions to Mars and beyond. *Futures* 115C, 102489.

Tännsjö, T. 1998. *Hedonistic utilitarianism*. Edinburgh University Press.

Taylor, F. W. 2010. *The scientific exploration of Mars*. Cambridge University Press.

Toivonen, A. 2021. *Sustainable space tourism: An introduction*. Channel View Publications.

Torres, P. 2020. Can anti-natalists oppose human extinction? The harm-benefit asymmetry, person-uploading, and human enhancement. *South African Journal of Philosophy* 39 (3): 229–245. https://doi.org/10.1080/02580136.2020.1730051.

Tozzo, P., S. Zullo, and L. Caenazzo. 2020. Science runs and the debate brakes: Somatic gene-editing as a new tool for gender-specific medicine in Alzheimer's disease. *Brain Sciences* 10 (7): 421. https://doi.org/http://dx.doi.org/10.3390/brainsci10070421.

Traphagan, J. 2016. *Culture, science, and the search for life on other worlds*. Springer.

Travis, J. 2015. Making the cut: CRISPR genome-editing technology shows its power. *Science* 350 (6267): 1456.

Turocy, J., E. Y. Adashi, and D. Egli. 2021. Heritable human genome editing: Research progress, ethical considerations, and hurdles to clinical practice. *Cell* 184: 1561–1574.

United States Senate. 2017. *Gene editing technology: Innovation and impact; Hearing of the Committee on Health, Education, Labor, and Pensions*. United States Senate, One Hundred

Fifteenth Congress, first session, on examining gene editing technology, focusing on innovation and impY 4.L 11/4:S.HRG.115-663

Unruh, C. F. 2021. The strings attached to bringing future generations into existence. *Journal of Applied Philosophy* 38: 857–869. https://doi.org/10.1111/japp.12532.

Van Pelt, M. 2005. *Space tourism: adventures in Earth orbit and beyond.* Copernicus Books.

Van Slyke, J., and K. Szocik. 2020. Sexual selection and religion: Can the evolution of religion be explained in terms of mating strategies? *Archive for the Psychology of Religion* 42 (1): 123–141.

Veatch, R. M., and L. K. Guidry- Grimes. 2020. *The basics of bioethics* (4th ed.). Routledge.

Vincent, N. A., and E. A. Jane. 2019. Parental responsibility and gene editing. In E. Parens and J. Johnston (Eds.), *Human flourishing in an age of gene editing* (pp. 126–139). Oxford University Press.

Visser, S. L. 2003. The soldier and autonomy. In Thomas E. Beam and Linette R. Sparacino (Eds), *Military medical ethics* (Vol. 1, pp. 251–266). Office of the Surgeon General at TMM Publications. Borden Institute.

Walker, Margareth U. 1998. *Moral understandings: A feminist study in ethics.* Routledge.

Walter, M. 1999. *The search for life on Mars.* Foreword by Paul Davies. Perseus Books.

Walter, V., J. Anomaly, N. Agar, P. Singer, D. S. Fleischman, and F. Minerva. 2021. Can "eugenics" be defended? *Monash Bioethics Review* 39: 60–67.

Wanjek, C. 2020. *How humans will settle the Moon, Mars, and beyond.* Harvard University Press.

Warwick, K. 2020. Superhuman enhancements via implants: Beyond the human mind. *Philosophies* 5: 14. https://doi.org/10.3390/philosophies5030014.

Weintraub, D. A. 2018. *Life on Mars: What to know before we go.* Princeton University Press.

Wignall, P. B. 2019. *Extinction: A very short introduction.* Oxford University Press.

Wójtowicz, T., and K. Szocik. 2021. Democracy or what? Political system on the planet Mars after its colonization. *Technological Forecasting and Social Change* 166: 120619.

Wolpe, P. R. 2005. Dialogue: Bioethics in space. *Medical Ethics* 12 (1): 10–11.

Yashon, R. K., and M. R. Cummings. 2009. *Human genetics and society.* Brooks/Cole Cengage Learning.

Young, L. R., and J. P. Sutton (Eds.). 2021. *Handbook of bioastronautics.* Springer.

Zubrin, R. 2019. Why we earthlings should colonize Mars! *Theology and Science* 17 (3): 305–316.

Zubrin, R. M. (with R. Wagner). 1996. *The case for Mars: The plan to settle the red planet and why we must.* The Free Press.

Index